NUCLEAR
POWER
AND
SOCIAL
POWER

RICK
ECKSTEIN

Temple University Press

Philadelphia

To Monica, for generosity beyond expectations,
and to Emma, for relentless spirit during a tough first year

Temple University Press, Philadelphia 19122
Copyright © 1997 by Rick Eckstein
All rights reserved
Published 1997
Printed in the United States of America

♾ The paper used in this publication meets the requirements of the
American National Standard for Information Sciences—Permanence
of Paper for Printed Library Materials, ANSI Z39.48–1984

Text design by Erin New

Library of Congress Cataloging-in-Publication Data

Eckstein, Rick, 1960–
    Nuclear Power and social power / Rick Eckstein.
        p.   cm.
    Includes bibliographical references and index.
    ISBN 1–56639–485–6 (cloth: alk. paper). — ISBN 1–56639–486–4 (paper :
alk. paper)
    1. Nuclear industry—Government policy—United States—Citizen
participation   I. Title.
    HD9698.U52E25   1996
    333.792'94—dc20   96–23908

CONTENTS

ACKNOWLEDGMENTS

When you have worked on a project for as long as thirteen years, it is pretty hard to remember who really helped and who just happened to be hanging around. No matter. I am fortunate to have received priceless support from hundreds of people in dozens of places. Thanks first to my family for sticking with me for all or part of thirty-five years and for buying multiple copies of a book they will probably never read.

I offer my infinite gratitude to Cheryl Laz, Kevin Delaney, Debra Swoboda, Ed Royce, Sandra Hinson, and Monica McDermott, who provided invaluable intellectual and moral assistance at many stages of this project. Villanova students Marlene Schneider, Steve Albrecht, Melinda Kane, Nnenna Lynch, Brendan Walsh, Joe Kroart, Rebecca Schoenike, and Dean Sicoli also left their mark on this research through their labor and their intelligence. Other Stony Brook and Villanova students who passed through my classes since 1985 gave my academic life purpose and meaning.

Eight years of graduate school builds character as well as debt. My friends and teachers at SUNY, Stony Brook, kept me sane and focused (when possible) and even taught me a thing or two. Thanks to Michael Schwartz, Mark Granovetter, Richard Williams, David Halle, Michael Zweig, Lyle Hallowell, David Lee, Bob Boice, Wanda Olivera, Judy Thompson, Rosemarie Sciales, Carole Roland, Veronica Abjornson, Lee Clarke, Randy Simmons, Shan Nelson-Rowe, Milly Peña, Libby Chute, Patrick McGuire, Ray Maietta, Chris Vestuto, and Dom Chan. Special commendations to all my comrades in the Graduate Student Employees Union for fighting the good fight and offering the necessary distraction from academic work.

My friends and colleagues at Villanova University have provided tangible support and kept the faith (no pun intended) as I struggled to complete this book even as personal tragedy enveloped me. My appreciation to Bill Waegel, Father Kail Ellis, Bill Werpehowski, Barbara Wall, Tom Arvanites, Dave Leibig, Jack Doody, Dan Regan, Rachel Schaller, Charlotte Vent, Satya Pattnayak, Bob DeFina, Brian Jones, Rich Juliani, Miriam Vosburgh, and the Center for Peace and Justice Education.

None of this was possible without the inspiration of being an undergraduate at Marietta College. Dave Boyer and Larry Burnett remain to this day the greatest influences on my intellectual spirit. Important supporting roles were and continue to be played by Catherine Burns, Margaret Barker, Julie Haught, Marcy Dupay, Dawn "Delbert" Lausa, Amy (Bates) Van Hise, and Sara Shute.

I barely know Doris Braendel, my editor at Temple University Press, but I am willing to take her grandchildren to Chuck E. Cheese's in appreciation for her support. This was a terrible time for me to be finishing a book, but Doris never let me stop trying. Thanks also to copy editor Sue Gleason for unmixing my metaphors.

Finally, a few words will never do justice to the love and support I received from Monica Nicosia—spouse, confidante, proofreader, and favorite cell biologist. I was in the middle of this book when our (at the time) ten-month-old daughter, Emma Eckstein Nicosia, was diagnosed with a malignant tumor. Five months later, Emma is a very happy kid (with a very dead tumor), and this book is finished. Monica's herculean efforts and unprecedented selflessness were greatly responsible for both accomplishments. I owe her big time.

R. E.

LIST OF ACRONYMS

ACRS   Advisory Committee on Reactor Safety

AEC   Atomic Energy Commission

ASLB   Atomic Safety and Licensing Board

BIA   Business and Industry Association

CWIP   Construction While in Progress

FEMA   Federal Emergency Management Agency

JCAE   Joint Committee on Atomic Energy

LIA   Long Island Association

LILCO   Long Island Lighting Company

LIPA   Long Island Power Authority

NHPUC   New Hampshire Public Utilities Commission

NHY   New Hampshire Yankee

NRC   Nuclear Regulatory Commission

NYPSC   New York Public Service Commission

PSNH   Public Service Company of New Hampshire

PRDP   Power Reactor Development Program

RICO   Racketeer-influenced and Corrupt Organizations

SAPL   Seacoast Anti-Pollution League

TMI   Three Mile Island

CHRONOLOGY

A Concise History of Shoreham

Date	Event
Date	*Event*
November 1968	LILCO submits application to AEC for construction permit
September 1970	Construction begins, construction permit hearings begin
April 1973	AEC issues construction permit
March 1979	Three Mile Island accident
October 1980	NRC issues new guidelines for emergency evacuation
February 1983	Suffolk County withdraws from Shoreham evacuation plan
January 1984	LILCO executive reorganization
	LILCO witholds local property taxes
August 1984	LILCO restructures debt
May 1985	Suffolk County executive Cohalan shifts position on evacuation
September 1985	Hurricane Gloria hits Long Island

December 1985	NY PSC rules that LILCO was imprudent in Shoreham spending
January 1986	LILCO restructures debt
May 1986	Suffolk County sues LILCO for lying to NYPSC
January 1987	Long Island Power Authority created
October 1987	NRC institutes "realism clause" concerning emergency evacuation
September 1988	Federal Emergency Management Association approves evacuation plan
May 1988	LILCO and New York agree on original one-dollar deal to close Shoreham
February 1989	LILCO and New York agree on second one-dollar deal to close Shoreham
April 1989	NRC grants Shoreham full power license
June 1989	One-dollar deal finalized; Shoreham closes

A Concise History of Seabrook

Date	Event
March 1973	AEC receives construction permit application
July 1976	NRC issues construction permit
May 1978	New Hampshire Public Utilities Commission allows CWIP costs into rate base
March 1979	Three Mile Island accident
May 1979	New Hampshire legislature prohibits CWIP rule change

October 1980	NRC issues new guidelines for emergency evacuation
March 1981	Massachusetts officially opposes evacuation plan
August 1982	PSNH credit rating lowered to barely investment grade
February 1984	PSNH executive reorganization
June 1984	PSNH restructures debt; New Hampshire Yankee born
October 1987	NRC institutes "realism clause" concerning emergency evacuation
January 1988	PSNH enters Chapter 11 bankruptcy
December 1988	NRC approves evacuation plan
June 1990	NRC grants full power license

NUCLEAR

POWER

AND

SOCIAL

POWER

■

INTRODUCTION

In early 1990, the Long Island Lighting Company (LILCO) sold its $5.5 billion Shoreham nuclear power plant to New York State for a dollar. This completed, fully licensed reactor would never produce a single volt of commercial electricity—a totally unprecedented event in the United States. There was fierce opposition to the Shoreham plant by many different groups of people, some of which claimed the plant's closing as a great victory against nuclear power and corporate greed. Yet despite apparently clear winners and losers, the situation is far more complicated. The Long Island Lighting Company is more financially sound than ever, despite "losing" $5.49999 billion on its reactor. "Victorious" Long Islanders currently pay the highest electric rates in the country.

Later in 1990, the Seabrook (New Hampshire) nuclear plant received an operating license and began producing electricity commercially. Seabrook cost about the same as Shoreham, was built at roughly the same time, and faced stronger grassroots opposition. In addition, Public Service Company of New Hampshire (PSNH), Seabrook's largest owner, entered bankruptcy in 1988 and remained under Chapter 11 protection while the plant received its license. Despite Seabrook's opening, PSNH was taken over by Northeast Utilities of Connecticut. Seabrook lives and PSNH dies; Shoreham dies and LILCO thrives.

These cases defy simple explanations. In this book I explore the complex political, economic, and social factors that contributed to the very different outcomes at these seemingly similar reactors. Although this study offers a new perspective on nuclear power in the United States, it is not just about atomic energy. Instead, the story of Shoreham and Seabrook illustrates larger issues about prevailing definitions of economic growth and development, corporate behavior, corporate interaction with other organizations (such as banks), government regulation of businesses, and how ordinary people can and cannot affect the circumstances of their lives. The Shoreham and Seabrook stories illustrate the many faces and dimensions of social power.

There are several shortcomings of academic and nonacademic efforts to make sense out of Shoreham and Seabrook. First, there has been little attempt to supplement the descriptive richness of case studies with larger theoretical ideas and historical patterns. Studies of individual plants treat them as independent of other plants and dislocated from the historical context of nuclear energy. Second, more theoretically interesting research tends to make generalizations about nuclear power that ignore the interesting idiosyncracies of individual cases. Third, theoretically informed studies of nuclear power rarely consider organizations, such as corporations, banks, and government agencies, as a primary unit of analysis. Instead, they emphasize industries, states, and movements. Organizations are not ignored, but they are usually treated only in passing. Any organizational analysis usually focuses on company "mismanagement" which paints an overly simplistic picture of corporate behavior. In short, there has never been a theoretically informed comparison of two or more specific nuclear plants and the organizations that build, finance, and regulate them.

The Theoretical Engine

A common model of academic research begins by advocating the superiority of one theoretical school over all others and then finding some small segment of social reality that supports this proclamation. Such an approach tends to give the theories themselves a starring role while making social reality second banana—not unlike working endlessly on a car engine in your garage until it purrs and sings, regardless of the fact that it is not part of a vehicle. The engine has become an end in itself rather than a means to move a car and provide mobility. The person fine-tuning the engine might say to his or her partner, "I've really got that engine humming," to which the partner might reply, "Okay, then take it to the store and buy beer."

A clean and shiny engine sitting in the garage may be nice, but not much help if you are thirsty. A clean and shiny theory is also nice, but not necessarily helpful for making sense out of social reality. Conversely, an empirical study of social reality without any theoretical analysis is flat and uninteresting. *Nuclear Power and Social Power* is a theoretically powered empirical study; it uses theories to explain social phenomena rather than using social phenomena to "test" pet theories. Initially perplexed by Shoreham and Seabrook, I employed certain sociological theories to help me make sense of things rather than selecting events to demonstrate the prowess of my favorite theories.

Making sense out of Shoreham and Seabrook requires an eclectic theoretical engine. This theoretical engine has four main parts: (1) *ideological* conceptions of "proper" economic growth and development; (2) *institutional* relationships among businesses and between business and government that reflect these ideological conceptions; (3) *organizational and interorganizational behavior* patterned by these ideological and institutional circumstances; and (4) *individuals* who influence and are influenced by these more structural phenomena.

These general theoretical concepts help generate the specific ideas that explain the similarities and differences of Shoreham and Seabrook. First, the conventional definition of "proper" economic growth is equated with a centralized, corporatized, capitalistic system of production and distribution. Nuclear power is consistent with this "big is good" perspective, although there may at times be dissension from this particular growth ideology. Such dissension is unusual but very interesting — especially as it played out at Shoreham and Seabrook.

Second, government entities at every level combine with corporations in "growth coalitions" that reflect this conventional growth ideology and help subsidize the expansion and legitimation of corporate capitalism. The nuclear industry (almost totally corporate dominated) has been involved in such a coalition with the government (federal, state, and local), which has promoted civilian nuclear power plants and the private, corporate utilities that build them. Sometimes these coalitions are strong, sometimes they are weak. Their relative strength is an important empirical question that is central to understanding Shoreham and Seabrook.

Third, corporations such as electrical utilities do not operate in an organizational vacuum. Actions and decisions within one specific organization are patterned by those in a multitude of other organizations. Events at Shoreham and Seabrook were not necessarily rooted in the internal managerial decisions of LILCO and PSNH executives, but rather in the interorganizational relationships encompassing the utility. The most important interorganizational relationships affecting Shoreham and Seabrook were between the utilities and their creditors.

Finally, grassroots activists play an important part in the general history of nuclear power and the specific stories of Shoreham and Seabrook. This role is often exaggerated and romanticized, sometimes as a tribute to the vibrancy of American democracy. The Shoreham and Seabrook story offers a much less sentimental view of grassroots opposition. Despite popular rhetoric, regular folks (and their interests) were swept away by more powerful social forces (and their interests). Most importantly, grassroots opposition—what I call "people power"—cannot be viewed as some homogenous monolith that affects nuclear power and nuclear plants. The opposition at different plants may not look the same, act the same, or have the same efficacy. Grassroots opposition at specific plants must be located within a particular local political-economic context.

Each of these theoretical ideas has something to do with social power. Individuals or organizations that control certain structural and ideological resources can use them to satisfy their own interests at the expense of others. Social power can take many forms, some of them obvious and others more subtle. The stories of Shoreham and Seabrook uncover many subtle faces of social power. For instance, government regulations on nuclear power often reflect the conventional "good growth" paradigm, which favors centralized corporate capitalism. Yet the specific corporations (and their members) that benefit from these decisions may have taken no part in them. Instead, it is as if social institutions are automatically programmed to serve specific parochial interests. Again and again in the Shoreham and Seabrook sagas, decisions that served specific class and organizational interests were justified as being in everyone's best interests.

A Brief Overview

This analysis began with very brief histories of the Shoreham and Seabrook nuclear plants. These histories were brief for two reasons. First, later chapters will provide more details. Second, I do not want to fall into a "chronology trap" that emphasizes names, dates, and periods. I want to demonstrate how interesting theoretical ideas can power an empirical analysis. Chapter 1 begins with a critique of the conventional explanations for what happened at Shoreham and Seabrook, and develops an alternative framework driven by the theoretical engine I mentioned earlier.

Chapter 2 applies this framework to the history of civilian nuclear power in the United States. There are already many excellent studies about nuclear power's development, and this chapter will offer little new information. What is important here is using this different framework to think about the same historical events. Chapter 2 sets the context for Shoreham and Seabrook.

Chapters 3, 4, and 5 apply the same ideas specifically to Shoreham and Seabrook. This discussion challenges three conventional explanations for Shoreham's closing and Seabrook's problems: regulatory hostility, managerial incompetence, and the strength of "people power." Chapter 3 examines one piece of the mosaic of institutional growth coalitions: the relationship between government regulatory agencies and the nuclear utilities constructing Shoreham and Seabrook. At the federal level, the Nuclear Regulatory Commission, or NRC (formerly the Atomic Energy Commission, AEC) was the most important agency affecting LILCO and PSNH. The New York Public Service Commission (NYPSC) and New Hampshire Public Utilities Commission (NHPUC) also played critical roles in both cases. Decisions by these federal and state regulators did not "cause" the problems faced by LILCO and PSNH. In fact, both plants were being built within a friendly and supportive regulatory system, even if that system was perhaps not as friendly as it once had been.

Chapter 4 calls attention to the interorganizational relationships between the utilities and their creditors. Nuclear plants are very expensive, and both LILCO and PSNH were unable to finance Shoreham and Seabrook strictly out of internal funds. This capital shortage led them to external capital markets, where money was available but accompanied by significant costs, caveats, and contingencies. As this dependence on external capital grew, the financial community became much more powerful, while LILCO and PSNH managers became less powerful. Blaming utility managers for the complex events at Shoreham and Seabrook ignores this fascinating interorganizational element.

Chapter 5 integrates many of the four main theoretical ideas to explore the local political economies in which Shoreham and Seabrook were built and the importance of "people power" at each plant. Here we find the key factor contributing to the very different outcomes at each plant. In short, Shoreham's unprecedented problems and closure were rooted in Long Island's weak ideological consensus and institutional coalitions around "proper" economic growth and development. Competing growth paradigms provided fertile ground for the unprecedented opposition of a local government (and eventually a state government) to a nuclear plant and exaggerated the

efficacy of grassroots opposition. The New Hampshire political economy displayed much more ideological and institutional cohesion
around the dominant "big is better" growth paradigm. Seabrook
faced even stronger opposition by unofficial people power but faced
few important "official" adversaries. This difference shows why it is
important not to overgeneralize about the opposition to nuclear
power.

The final chapter considers some of the implications of this research, especially on the distribution of power and risk within society and how this power distribution affects the way we study and think
about democracy. Very little in the Shoreham and Seabrook stories
testifies to the ability of regular people to influence the decisions that
affect their lives. Although the stories ended very differently, the
same groups, organizations, and classes generally won and lost.

Studying Shoreham and Seabrook

The eclectic theoretical engine driving this study is matched by an
equally eclectic methodology. For Shoreham, the bulk of the information comes from the popular and business press, with secondary
support from archival material, interviews, and observation. As Mc-
Caffrey (1991; 23) notes, Long Island's *Newsday* provided excellent
coverage of Shoreham, frequently including detailed interviews with
important players and excerpts from official documents. Although in
the best of all possible worlds I would have preferred scouring the
archives and conducting these interviews myself, that was neither
possible nor necessary. For instance, *Newsday* often presented ver-
batim official NRC rulings, which would later end up in its official
repository at the Shoreham–Wading River Public Library. I spent
many hours in this repository and found little of importance beyond
what was already printed in the papers. I did conduct some personal
interviews but found many key players unwilling or hesitant to discuss Shoreham on the record. Fortunately, these people had been

less shy about talking with the media as the Shoreham saga was un-folding.

Ironically, *Newsday* was an editorial champion of Shoreham and one of the few Long Island corporations that avidly supported the one-dollar deal that closed the plant (see Chapter 5). However, their editorial position in no way affected their substantive reporting on Shoreham. In fact, some people on Long Island wondered if *Newsday*'s editorial board was reading the stories printed in their own paper. Despite their accuracy and thoroughness, the newspaper reports were by no means objective, for the stories generally reflected conventional analytical biases. For example, because journalists believed that LILCO mismanagement was a key issue, many stories emphasized it at the expense of other possibly salient factors. Thus, I have basically drawn different conclusions from the data accumulated by the popular and business presses.

The popular and business presses also provided a great deal of information on Seabrook. In this case, the *Boston Globe* and *Manchester Union-Leader* were the papers of record, although they took antithetical views toward Seabrook. The Boston paper "opposed" the plant, whereas the largest paper in New Hampshire "supported" the plant and ridiculed its opponents. The Exeter (New Hampshire) Public Library is the repository for all AEC/NRC documents about Seabrook, and I spent many hours combing its illogically annotated files (which were *not* devised by the local librarians, who used all their cryptographic wits to guide me along). This particular part of the research was assisted by two monographs on Seabrook: James Stever's *Seabrook and the Nuclear Regulatory Commission* and William Bedford's *Seabrook Station*. Stever and Bedford had already done a great deal of archival legwork, but I did find some new items of interest. Bedford's book also provided priceless interviews with important people and excerpts from small local newspapers. I had no possible access to these sources. I have a different take on things than Mr. Bedford has, but could not have reached these distinct conclusions without his help.

I have broken some new empirical ground on Seabrook by analyzing U.S. Bankruptcy Court documents about PSNH's Chapter 11
case. This was another painstaking journey through a labyrinth of
poorly filed and untitled material, through no fault of the friendly
and helpful folks at Ray's Quick Copy in Manchester, New Hampshire—the unofficial repository of all PSNH bankruptcy documents.
There may still be kernels of truth piled somewhere in Ray's base-
ment awaiting discovery by someone with a few weeks of idle time. I
was also lucky to have interviewed Ted Feigenbaum and Ed Brown,
who in 1992 were president and past president, respectively, of New
Hampshire Yankee, the titular owner of Seabrook since 1985 (see
Chapter 4). Mr. Feigenbaum and Mr. Brown provided a plethora of
important information and invaluable insights unavailable elsewhere and, to my knowledge, rarely solicited.

Why Study Shoreham and Seabrook?

The implications of this research may seem depressing to those who
believe that a truly democratic society is desirable and achievable.
But I would suggest that a study like this can be liberating because it
exposes some of the hidden social processes that constrain our ability to make history under conditions of our own choosing. Exploring
and understanding these processes, rather than just looking at their
outcomes, opens new doors for understanding how an individual's
personal fulfillment and frustration are linked with more complex social issues. This book reflects a commitment to C. Wright Mills's
idea of a "sociological imagination" that seeks to link personal troubles with public issues. For Mills, the single greatest obstacle to a fair
and just society was the unequal and unfair distribution of social
power. He called on sociologists to confront these injustices by
clearly illuminating them for regular people. Once regular folks better understood complex social arrangements, they would be more
able to make personal history in fulfilling ways.

I found the Shoreham case to be a perfect topic for doing interesting and relevant sociological research. I was born and raised on Long Island and, after a four-year stint at Marietta (Ohio) College, returned home for graduate school at SUNY, Stony Brook. Consequently, I was on Long Island throughout the eighties as the Shoreham saga unfolded. I discovered during many years of research that few people understood the complexity of the situation or recognized that the "Shoreham controversy" was linked to more general social issues of wealth, power, and inequality.

Studying Shoreham also proved to be important pedagogically. Since most Stony Brook students hail from Long Island, they were fascinated by a detailed examination of Shoreham in that it demonstrated how sociological analysis is more than just a bunch of fancy terms for everyday phenomena. Even at Villanova University, where I have taught since 1990, the students seem to appreciate these insights into complex matters that do affect their lives—in this case, their electric bills. My students are a continual source of feedback and inspiration. Their enthusiasm convinced me that this topic was worth pursuing and might even make some small difference beyond the captive audience of the classroom and the standard academic journals.

Seabrook entered the picture because these complex social patterns are more interesting (and believable) when they happen in more than one place. These two cases cried out for a comparative study because they had so much in common yet ended so differently. The explanations in the newspapers and on television (and in some academic works) either just scratched the surface or ignored important issues. Nevertheless, Seabrook plays a supporting role in this study; Shoreham has the lead. The reason for this is simple: Shoreham is far more unusual and unique. Seabrook is interesting but is really just another nuclear plant that faced some opposition, cost lots of money, and eventually opened. The most important thing about Seabrook was PSNH's bankruptcy, an event that has been given a prominent part. The book, then, is empirically biased toward Shoreham. But the strength of the theoretical ideas depends on both cases.

Talking about nuclear power has an inherent problem: people want to know "which side" you are on. I have tried to sidestep the standard pronuclear/antinuclear debate, which usually focuses on technological safety. Instead, I am more interested in social power and how people and organizations with more of it can constrain the behavior of those with less of it. On this issue, the book does take sides: it is "prodemocracy" in its classic participatory sense. I think it is wrong when relatively powerless people and organizations subsidize the interests of relatively powerful people and organizations— either with or without overt actions by the powerful actors. Such an ethical foundation should make this study interesting beyond the typical polemics about nuclear power.[1]

I chatted recently with an engineer friend who wondered why there was so much opposition to nuclear power in the United States. We talked for a long time about nuclear plant financing and how countries like France and Sweden were scaling back their nuclear capacities for purely economic reasons. My friend was both stunned and fascinated that nontechnological matters played such a prominent part in what he had always thought was a purely technological issue. I would argue that issues of social power underlie almost everything that we experience on a daily basis. The difficulty of identifying the many faces of social power is all the more reason to study it.

CHAPTER ONE

THE MANY FACES

OF SOCIAL POWER

■

Explanations for the unusual events at Shoreham and Seabrook are remarkably similar despite the very different outcomes of each case. Three basic themes are almost always present: a hostile regulatory environment, incompetence by LILCO and PSNH management, and grassroots opposition. These factors are supposedly responsible for the massive delays and cost overruns at both plants, and for the unprecedented closing of Shoreham.

These common explanations are not so much incorrect as they are inadequate. Government policies certainly played a part in these stories, but they were not tantamount to regulatory hostility. Managers could have (and perhaps should have) made different decisions, but claiming so overlooks the larger organizational context in which executive decisions were made. Grassroots opposition played an important role but was by no means a sufficient cause of the problems that plagued both plants. There was much more going on at Shoreham and Seabrook than these three suppositions permit us to see. It is absolutely necessary to examine the regulatory frameworks, managerial decisions, and "people power" in both cases, but these are not adequate explanations for what happened at Shoreham and Seabrook.

More theoretically sophisticated arguments also fall short in explaining what happened at Shoreham and Seabrook. For example, Jim Jasper's (1990) outstanding *Nuclear Politics* compares the divergent trends in the American, French, and Swedish civilian nuclear power programs. Jasper argues that these divergent trends are best explained by looking at the dynamics of state policy making (1990, 4). John Campbell's (1988) equally admirable *The Collapse of an Industry* examines how different industrial policies (broadly defined) patterned divergent civilian programs in the United States, France, Sweden, and West Germany. Both studies use the nation-state as their primary unit of analysis and generate important insights about nuclear power in general. But these national generalizations are incapable of addressing the vastly diverse nuclear circumstances within a particular country. This does not invalidate these theories; it only means that they are not useful for framing a discussion of nuclear power that is more interested in organizations than in nation-states. Shoreham and Seabrook cannot simply be plugged into these paradigms.

The Many Faces of Social Power

Shoreham and Seabrook are as much about social power as about nuclear power. We usually think about power as some *identifiable* individual or organization influencing, constraining, or controlling other individuals or organizations. But power is often hard to identify. What we might see as balanced relationships between and among social actors may actually be unequal relationships. Even when acknowledging social power, we often fail to wonder why these unequal relationships exist. We assume that inequality is natural and that there must be some good reason why some social actors have power over others.

Social power can take many shapes and forms and operate in a variety of venues. Steven Lukes (1974) believes that social power operates in three different arenas, or what he terms "dimensions." The first dimension encompasses normal, easily observable political arenas where various interests present their cases and fair, visible political mechanisms decide what should be done. Second-dimensional power adds another layer, as less conspicuous actions (such as agenda setting) explicitly keep certain alternatives from being considered by decision makers. Here, power differentials become important as certain people and groups have better access to material and ideological resources that can bias decision making (or nondecision making) toward their interests. Although this is a slightly more sophisticated approach to understanding power, it still insists that the action is observable, if not always easy to see.

The third dimension of power is truly sociological since it considers more subtle, less observable *social* forces that prevent certain alternatives from even being considered without the need for direct agency by those benefiting from certain decisions. As Lukes puts it, "Is it not the supreme and most insidious exercise of power to prevent people, to whatever degree, from having grievances by shaping their perceptions, cognitions, and preferences in such a way that they accept their role in the existing order of things, either because they can see or imagine no alternative to it, or because they see it as natural and unchanging " (1974, 24). Although this dimension of power is harder to see, it is by no means invisible. It often operates in the realm of ideology and is very similar to what Gramsci (1971) calls "hegemony." This subtle hegemony can be traced by identifying the interests of different individuals, organizations, and classes and whether or not specific social actions serve (or do not serve) those interests. It may be that certain people, groups, and even ideas emerge more "victorious" than others. At Shoreham and Seabrook, social power operated in all three dimensions, although the second and third are often understated or overlooked.

Robert Alford and Roger Friedland (1975, 1985) see power similarly labeling the dimensions "situational," "structural," and "systemic." Systemic power, like the third dimension of power, transcends the behavioral/participatory limitations of more conventional perspectives on power. This view maintains that the power enjoyed by certain interests is not necessarily predicated on their actual participation (although that is certainly possible). Instead, social relationships have been structured in a way to serve certain interests without agents of these interests having to get involved. Corporate interests are systemically served by "political" decision making, without any overt participation by benefiting corporations.

> Given their frequently superior financial resources, especially when compared to cities and states, and their locational flexibility, corporations present a constant constraint. The absence of corporate participation in political decision making does not indicate the extent of corporate systemic power. To the extent that growth and thus fiscal capacity depend on continued private control over investment . . . decisions, issues will not be raised or policies legislated that impinge upon that control. (Alford and Friedland 1975, 447–448)

Although third dimensional power and systemic power are practically synonymous, I will talk only about third-dimensional power while discussing Shoreham and Seabrook, mostly to avoid semantic confusion. This will allow me to use the term *structural* more freely when referring to theoretical ideas that go beyond a focus on individual people. For example, "mismanagement" and "people power" are individualistic, first dimensional explanations for the events at Shoreham and Seabrook because they see easily observable managers and voters as the key players. My alternative analytical framework is more "structural" because it transcends this emphasis on individuals and is more concerned with second-dimensional power (such as the power of banks to set intercorporate agendas) and third-dimensional power (such as the subtle strength of growth coalitions and growth ideologies).

A third-dimensional approach to Shoreham and Seabrook goes a little beyond the ideas I discussed above. It is not corporations but corporate capitalism that receives the spoils of third-dimensional power, even when corporations or corporate capitalists are not directly involved. In other words, it is the *logic* of private—primarily corporate—control over investment and accumulation decisions that is the consistent winner in a centralized, corporatized, capitalist political economy. Individual capitalists and corporations may or may not benefit from all social decisions, but corporations and capitalists as a whole generally have their interests served. For example, both the one-dollar deal at Shoreham and PSNH's bankruptcy left a trail of battered companies and capitalist class members. But these events only strengthened the logic of centralized, corporate capitalism by quietly steamrolling alternatives that challenged whether large, capitalist corporations are the best generators and distributors of electricity. It is not so much that identifiable individuals and organizations overtly suppressed alternatives as that these alternatives never made it to the discussion table in the first place. Subtle social forces sustaining the logic of centralized, corporate capitalism prevented such options from ever becoming a possibility despite an occasional token appearance.

Social actors will often rationalize and legitimate the logic of corporate capitalism even when they stand to gain nothing, because they think it is the right thing to do. Ideologies and actions that support the logic of corporate capitalism are defined as natural, whereas those that challenge this logic are seen as radical or abnormal. For instance, most people do not question reducing corporate taxes if a company stays in a region and creates jobs. Even when a corporation leaves town anyway (or perhaps demands a new, publicly financed stadium) few people question the "trickle-down" logic although they might lambaste the particular company. On the other hand, there is likely to be widespread opposition to programs offering tax incentives and public subsidies to poor people that might allow them to start businesses and "create jobs."

Robert Goodman (1986) provides an excellent example of just this sort of third-dimensional power in his study of local government subsidies to corporations. Not only do municipalities offer tremendous economic incentives to attract big businesses, but competition with other municipalities over the same companies leads to even greater subsidies. Ironically, once a business accepts these gratuities, the winning locality becomes almost automatically disadvantaged as other municipalities try to lure the business away with still more exotic deals. This illustrates ideological and structural components of third-dimensional power. Ideologically, it has become socially internalized (or automatically accepted as correct) that providing incentives to business will stimulate local economic activity to the entire community's benefit. This view is shared by both social elites and the masses—whose taxes will surely rise to compensate for this corporate subsidy. Structurally, the benefiting corporations do not have to ask or lobby for these tax breaks (although they often do), because everyone in the community assumes this is the best and most natural way to stimulate local economic growth.

Nuclear power is a wonderful avenue for studying the many faces of social power (especially the third dimension) because it is practically synonymous with centralized, corporate capitalism. Nuclear power epitomizes a "bigger is better" attitude whereby fewer plants generate more power, and whereby large corporations tackle the great financial and administrative challenges posed by these big reactors. Nuclear power is seen as a "natural" human progression. Not all individuals or organizations have concurred with this, but objectors always found themselves paddling against powerful third-dimensional tides. When some dissenters managed to move, the currents always found ways to tip the boat.

Studying Social Power

This diverse framework of social power informs my specific explanations for Shoreham and Seabrook. As I discussed in the Introduction, these explanations examine ideological, institutional, organizational, and individual factors regarding these two specific nuclear plants and their place in the overall history of American civilian nuclear power. Understanding Shoreham and Seabrook requires a broad definition of social power and the willingness to look beyond easily observable individual and organizational behavior.

Ideological Growth Consensus

In the United States, there is a certain normative agreement that economic growth and development is a good thing. We even agree on the specific form economic growth should take. Many of us probably remember elementary school projects in which we designed a futuristic community. We probably generated ideas that had razzle-dazzle technology at their core and an increasingly centralized political economy. Nobody in this community worked, yet every material desire was miraculously satisfied. Rarely did these future communities focus on a decentralized mode of production with less commodified labor and with fulfillment rooted in human labor rather than leisure.

In the conventional American ideology, few people challenge the holiness of "economic growth and development." Rarely discussed are precisely what is meant by "normal" growth and who benefits and loses from it. On the whole, this standard construction of "good" economic growth is urban and corporate — it is a centralized conception. Less centralized modes of production do not always share the same vision of growth and are often viewed pejoratively because they diverge from a corporate model. Phrases such as "rural hicks" and "country bumpkins" are more than just charming terms of a simpler, quieter life; they abet negative connotations that people living this way are not quite normal. This way of life may even be an anachronism to the normal course of economic (and even human) progress.

Nuclear power fits securely within this image of good economic growth and development because it is big, centralized, and corporate. American civilian nuclear plants have been built within a compatible ideological atmosphere, both nationally and locally. This does not necessitate an ideological consensus about the benefits of nuclear reactors. Instead, the relative strength or weakness of any growth consensus is an empirical question that may generate very different stories in different cases. On Long Island, for instance, there was very little consensus about what ideological paradigm best stimulated economic growth and development. Some individuals and organizations favored the conventional centralization paradigm of which the Shoreham plant was a crown jewel. Others argued for a softer approach focusing on tourism, small businesses, and environmental resources. Long Island was unusual because of the relative strength of this unconventional growth paradigm. Southern New Hampshire was far more convinced that what was best for corporate capitalism was best for everybody. There was some dissent from this view, but it never threatened the overall growth consensus. Thus Seabrook was built in a much more supportive local ideological climate than Shoreham.

Institutional Growth Coalitions

It is impossible to discuss these ideological issues apart from the structural mechanisms that articulate and implement them. My idea of institutional growth coalitions extends Molotch's (1976) discussion of "growth machines" and Logan and Molotch's (1987) idea of "the political economy of place": "The political and economic essence of virtually any given locality . . . is growth. The desire for growth provides the key operative motivation towards consensus for members of politically mobilized local elites, however split they might be on other issues. The very essence of locality is its operation as a growth machine" (Molotch 1976, 309–310). Most sociological research of growth machines takes for granted the existence of a local consensus on good economic growth and development[1] — mostly because these studies look at urban areas where the centralized vision of growth has long been dominant. Despite pockets of resistance in these cities to various quasipublic projects (such as mass transit), the prevailing ideology supports definitions of urban renewal (for example) that would raze working-class row houses and replace them with gargantuan glass office buildings.

But, as with growth ideologies, the existence and strength of growth coalitions should be treated as an empirical question. In fact, the *absence* of growth coalitions may be more interesting than their *presence.* Growth coalitions may also take different shapes. For example, Shoreham and LILCO benefited from a strong growth coalition including federal and state regulators like the NRC and New York Public Service Commission (NYPSC), but there was no corresponding local growth coalition. There was little agreement among Long Island businesses or between the local business community and the local government about how Shoreham should or should not fit into Long Island's future. In Chapter 5 I will argue that this absence of a strong local growth coalition (and its corresponding ideology) was the single most important difference between Shoreham and Seabrook.

Growth coalitions operate in all three dimensions of power. Sometimes coalitions operate openly, and sometimes they operate invisibly. The third-dimensional element of growth coalitions is apparent when individuals, organizations, and classes that do not participate overtly in the coalition's actions nonetheless benefit from these actions. Third-dimensional power may or may not be preceded by more observable participation by the benefiting social actors. In either case, the dominant parties will have their interests served. These growth coalitions are led by federal, state, and local government agencies making policies that support and legitimate centralized, corporate capitalism (Friedland and Palmer, 1984).

Nuclear power's regulating framework illustrates the structural bias of government-directed growth coalitions. Contrary to much popular wisdom, federal and state regulators were in no way hostile to the civilian nuclear industry and public utilities generally, or to LILCO, Shoreham, PSNH, and Seabrook specifically. Chapters 2 and 3 will highlight the historical support for civilian nuclear power by the Atomic Energy Commission (AEC) and its successor, the Nuclear Regulatory Commission (NRC). Chapter 3 will also explore how state agencies responsible for regulating public utilities are active members of institutional growth coalitions that support centralized, corporate capitalism.[2] These regulators seem more interested in the economic health of the regulated utilities than in the economic health of utility customers. With some notable exceptions, decisions made by state and federal agencies consistently sustain the structural and ideological elements of the growth coalition, even though agency members do not benefit in any way. As with federal regulation, it is not necessary to belong to a benefiting group to believe that these groups should benefit. Regulatory agencies subscribe to the world view that general social interests are best served when the companies they oversee are healthy and profitable. New York's Public Service Commission and New Hampshire's Public Utilities Commission exuded this attitude toward LILCO and PSNH, although there were occasionally decisions that hurt the utilities and their organizational interests.

Occasional negative decisions in no way constituted a hostile regulatory climate which destroyed the American civilian nuclear power program. The *Wall Street Journal* often served as a barometer for this attitude, and its news stories and editorials consistently excoriated federal and state regulators for being "challenged by a bunch of college students free of classes for the summer and with nothing better to do than set up picket lines [near nuclear plants]" (*Wall Street Journal*, 5 July 1978, 8). After Three Mile Island, the *Journal* considered the NRC nothing more than a mutinous captain who abandoned ship just when the nuclear industry needed it most (*Wall Street Journal*, 22 September 1981, 26; see also Cohen 1990, 163). We will see in the next two chapters that this alleged regulatory hostility is more fancy than fact. Federal and state agencies were usually eager participants in the growth coalitions supporting nuclear plants and nuclear power.

Organizational Relationships

The conventional emphasis on Shoreham's and Seabrook's mismanagement is rooted deeply in the "managerialist" view of corporate behavior (see Glasberg and Schwartz 1983). This position gained prominence in the 1930s with the publication of Berle and Mean's

(1932) *The Modern Corporation and Private Property*. Soon there-
after, James Burnham (1941) expanded, in his words, this "revolutionary" perspective. More recently, Alfred Chandler (1977) has become a sort of dean of managerialism. This perspective is built on
three main propositions. First, corporations have become a dominant form of business organization. Second, stock ownership in these
corporations has become increasingly dispersed and has reduced the
economic power traditionally wielded by "robber barons" and their
families. Third, skilled professional managers have replaced tradi-
tional owners at the corporate helm.

As a result, the "visible hand of management" deftly guides the
modern corporation toward progress and increased efficiency and
away from profiteering. Only professional managers are capable of
this new thinking, because, unlike owners who want to maximize
profits at any cost, managers are willing to "profit satisfice" in the
name of efficiency (see Cyert and March 1956). The continual dispersion of corporate stock allows managers to go about their professional business unfettered by personal or family greed. Business philosophy now revolves around the efficient survival of the corporation
rather than the accumulation of wealth. This leads to a more "soulful" corporation which, with profit maximization a thing of the past,
allows firms to better look after the wishes and desires of both their
employees and the society they serve (Gordon 1945; Kaysen 1957;
Marris 1964; Rockefeller 1964).

Managerialist premises are both logically suspect and difficult to
measure empirically. For example, management is supposedly concerned with efficiency rather than with profit maximization. But how
is efficiency evaluated outside of the tautological explanation that,
because a firm survived, it must be efficient? Can firms be inefficient
and still survive, or efficient and still die? LILCO managers were very
inefficient in opening Shoreham, yet their organization thrived.
PSNH managers were very efficient in seeing that Seabrook opened,
yet their organization perished. Ironically, these apparently efficient
PSNH managers led their company to bankruptcy—the epitome of

executive incompetence and inefficiency. Similar "incompetence" may have contributed to the very efficient opening of Seabrook. These confusing outcomes at Shoreham, Seabrook, LILCO, and PSNH cannot be explained by a simplistic relationship between efficiency and organizational survival.

Managerialism also ignores the fact that profits are still the most common measuring stick of efficiency. In addition, dispersion of stock ownership may not necessarily imply a more democratic, soulful corporation. Owning as little as 5 percent of a company's stock may offer significant control of a firm's behavior (Zeitlin 1978). The dispersion of stock is also empirically suspect. Only 31 percent of U.S. citizens owned any stock between 1987 and 1989 (*Business Week*, 26 June 1989). This alleged separation of corporate ownership and control also overlooks the increasingly important ownership role played by large institutional investors who buy and sell tremendous amounts of stock. Institutional investors have very different goals than individual investors, and very different amounts of power with which to realize their objectives (Eckstein and Delaney 1993).

Managerialism's most important shortcoming is its insistence that only the internal structure of organizations are important. This focus on internal management eschews the important interorganizational connections among various firms and ignores the power some organizations have over others. Often, a company's managers do not write the menus from which they select their organizational strategy. Instead, their options and alternatives may be constrained by extraorganizational forces over which they have little control. This unequal distribution of organizational power may be hard to see, but it should not be ignored. LILCO and PSNH were not acting in organizational vacuums while building Shoreham and Seabrook.

Finance hegemony theory (Mintz and Schwartz 1985) offers an antidote to this emphasis on mismanagement and opens up new doors for understanding Shoreham and Seabrook. From this perspective, financial institutions (usually large commercial banks and insurance companies) have an asymmetrical amount of power over other corporations, although they do not wield absolute control. This power differs based on the availability and affordability of capital throughout the corporate community. The power of financiers increases greatly in times of tight money. Financial institutions *constrain* the alternatives available to nonfinancial business and non-business organizations. This allows for some organizational discretion despite the constriction of possible choices by the financiers.

Financial hegemony works in three basic ways. First, the financial community's willingness to lend money (and at what price) both creates and eliminates alternatives for capital-hungry organizations. Companies often do not have the internal capital necessary to pursue their organizational goals. Second, financial organizations are often (although not always) the largest corporate institutional investors. This gives financiers the potential to influence corporate behavior either by "voting their shares" or by threatening to "dump" large quantities of stock (see Glasberg 1981; Eckstein and Delaney 1993). Third, banks have considerable access to knowledge about the corporate community, which they can use to influence their lending decisions and institutional stock activity (see Useem 1981). Banks acquire this information through their copious and strategic membership on other companies' boards of directors (Koenig and Goegel 1981; Mintz and Schwartz 1981; Mizruchi 1982). Financial hegemony worked in all three ways at Shoreham and Seabrook, although the first two are more important than the third.

This study of Shoreham and Seabrook expands finance hegemony theory to include how general external capital flows constrain managerial autonomy. Banks are not the only source of external capital for corporations. Companies often raise money by selling stocks or bonds. These methods of generating external capital are often preferred because they cost less and come with fewer contingencies. Bank loans may come due in a year or two, whereas most bonds do not mature for ten years or more. Financial organizations may underwrite a bond or stock sale, but this provides far less power than direct loans or institutional investment.[3] Much more important than this underwriting are the ratings that evaluate the relative strength of a company's bond or stock offerings. These ratings are part of an intricate web of factors affecting the affordability and accessibility of external capital for organizations in need of cash.

There has been very little systematic analysis of how bond and stock ratings get established. Ratings are usually treated as objective indicators of the speculative risk associated with investing in a specific company. The business press typically cites these ratings as if everyone knew about and agreed on how they were generated. Some investors put little value in these ratings and point to the frequently contradictory assessments by the two major raters: Moody's and Standard & Poor's (Lamb and Rappaport 1980, 51–52). Nevertheless, the general investment culture treats these ratings as accurate indicators of stock and bond reliability.

The assumed independence of ratings analysts may also be more theory than fact. In 1989, Donald Trump successfully pressured an agency to fire an analyst who had downgraded a recent Trump bond issue.[4] Some rating agencies have instituted a system whereby analysts receive a commission based on the accuracy of their recommendations (*Business Week*, 18 December 1989). This could ultimately lead to "rating inflation," as successful bond sales, typically correlated with good ratings, become a source of revenue for analysts. As with other leading economic indicators, a credit rating's ability to affect people's behavior is not necessarily caused by the rating's accurate reflection of reality. If people believe the ratings reflect reality, this belief will be real in its consequences. In many cases, with the advantage of hindsight, the predictions and ideas of these analysts were dead wrong. But their opinions did alter individual and organizational behavior, although not always in ways we might predict.

For example, lowered ratings are normally associated with an ailing company, raising doubts in investors' minds that they will make profits—much less recover their principal—by putting capital into the company. To balance such reservations, a company normally raises bond interest rates or offers a higher dividend on its stock. Thus increased risk creates the possibility of increased profit, just as introductory economics textbooks would predict. However, not all investors buy stocks and bonds speculatively or want to risk their capital. When LILCO's ratings began plummeting in the late seventies and early eighties, and interest rates therefore increased, LILCO actually became a more attractive risk for some investors; namely, the large institutional investors who could afford to speculate. Meanwhile, individual investors who were more concerned with steady dividend payments found this source of income cut off when LILCO and PSNH were forced by the banks to suspend their dividends.

There is also an interesting dialectic between this external capital and the internal capital that utilities generate through electricity rates. State utility commissions must approve rate increases, and they often use credit ratings to assess a company's health. Meanwhile, the keepers of external capital use levels of internal capital as one measurement of a utility's potential risk. Overseers of internal and external capital often make money available to utilities only if the other party first sends some sort of acceptable signal. Banks and ratings agencies often demand "rate relief" before lending money or raising ratings, but utility regulators want to see better credit ratings before granting rate relief. This cycle is exaggerated by the high cost of nuclear reactors. LILCO and PSNH were deeply enmeshed in this highly complex web of capital flows, which greatly constrained their managerial autonomy. Utility executives were not powerless, but they did not have complete control over available organizational choices.

Individual Opposition

During graduate school, whenever I launched into a sermon on structural constraint, one friend would always ask me, "Where are the people?" This is a legitimate question, because sociologists, frequently consumed with organizations, institutions, and ideologies, forget that human agency is an integral part of social structure.

Groups are not merely the sum of their individuals and individuals are not totally subservient to larger group dynamics. Individual action is important to the Shoreham and Seabrook stories, but it is still necessary to locate this action within its structural parameters. People who opposed these nuclear plants made history but not under conditions of their own choosing. The alternatives and options available to individual opponents of the plants were highly constrained. I hope neither to glorify nor to ridicule the importance of "people power" at Shoreham and Seabrook, but to more soberly assess the social forces that prevent regular people from making history more to their liking.

Individual opposition was a necessary but not sufficient cause of the various delays and idiosyncracies at Shoreham and Seabrook, and of the decline of civilian nuclear power in the United States. This opposition made a big difference, but it would not have been very effective without these structural elements. Shoreham probably would have opened without grassroots opposition. But this opposition by itself did not close the plant. Seabrook would have opened sooner without grassroots opposition, but this opposition alone never jeopardized the plant's completion.

Instead, it was the unprecedented "official" government opposition to Shoreham and Seabrook that had the greatest effect on each plant. Grassroots activity often sustained these official opponents, but it did not cause government opposition. At Shoreham, the Suffolk County government offered the most instrumental official opposition to Shoreham. New York State at first played a minor opposing role but supplanted Suffolk County in the late eighties. Seabrook's primary official opponent was the state of Massachusetts, which was part of the NRC-mandated emergency evacuation zone around the plant. There was also some opposition from townships and counties near Seabrook, but they never put up significant obstacles. Long Island and New Hampshire differed greatly in the strength and efficacy of this official opposition.

These different local dynamics between official and unofficial opposition are intertwined with some of the larger structural issues already discussed. The relative strength of opposition to Shoreham and Seabrook was directly related to the strength of the local growth coalitions and growth paradigms. Shoreham's unusually strong opposition was fueled by the absence of any significant local growth coalition either among corporations or between the corporate community and the local government. In the absence of any such coalition, grassroots activism assumed greater influence because it did not have to fight city hall or the corporate community; neither one was involved in any powerful growth coalition. Ironically, Seabrook's grassroots opposition was much more formidable than Shoreham's, but it was operating in a far less sympathetic environment. Seabrook's two main grassroots opponents, the Clamshell Alliance and the Seacoast Anti-Pollution League, were among the most creative and respected antinuclear groups in the United States. If "people power" were really sufficient to close a nuclear plant, it would have happened at Seabrook.

Even in the absence of a strong local growth machine, the alternatives available to Long Islanders were severely constrained. One reason for this was the ascension of an external growth machine—at the state level—which was able to more coherently create and eliminate certain possibilities consistent with a paradigm favorable to corporate capitalism. In the end, Long Islanders were offered two, and only two, choices: Shoreham opens or Shoreham closes on the condition that LILCO be given all the support it needs to return to financial health, including the highest electricity rates in the country. Other options had been on the table at one point or another, including the possibility of a consumer-owned utility. However, the emergence of this powerful state growth coalition swept such alternatives aside as "unrealistic" or "unobtainable."

My goal here is not to demean very significant actions by grassroots opponents. Rather, I hope to point out all the covert structural barriers that stand in the way of average people making history. Signing petitions, voting people out of office, canvassing neighborhoods, and taking the high moral ground simply may not be sufficient to achieve certain goals. The distribution of social power is usually inconsistent with the wishes, desires, and expectations of ordinary people who want to maintain some control over their lives. If democracy means that you can vote for competing political elites who offer slightly different views, then Long Island and southern New Hampshire were very democratic places. If, however, democracy means that people who are affected by decisions have the knowledge and ability to influence these decisions, then neither place was democratic at all.

Conclusion

Shoreham offers perhaps the most interesting case study in the history of American civilian nuclear power. Seabrook provides the perfect contrast. Together they allow for a more absorbing discussion of why certain nuclear plants have opened and others have not. This discussion can and should go beyond more prosaic explanations which rely on easily observable activities in normally observed decision-making arenas. There is often much more going on than we can see at first glance. The theoretical framework I outlined in this chapter offers a more compelling explanation for these two cases and their relationship with the American civilian nuclear power program. It also offers a chance to examine the many faces of social power and who gains and loses when power is exercised. The result casts a gloomy shadow over the viability of political and economic democracy in the United States.

CHAPTER TWO

THE HISTORICAL

PICTURE

■

Shoreham and Seabrook are just two pieces of a puzzle that has been under construction since the 1940s. If we want to understand how ideological definitions of economic growth, institutional business-government relationships, organizational financing, and individual opposition affected Shoreham and Seabrook, we also need to see how they influenced the overall development of civilian nuclear power in the United States. The history presented here is by no means exhaustive, and the material is not new. Scholars have already produced dozens of excellent studies of nuclear power, and I see no need to repeat them.[1] This chapter shows how the social forces shaping Shoreham and Seabrook are not case specific; that they were part of a larger political economic dynamic composed of plants other than Shoreham and Seabrook and organizations other than LILCO and PSNH. My goal here is to employ the theoretical framework outlined in Chapter 1 to reexamine empirically the history of nuclear power.

Ideological Issues: Concepts of Economic Growth

Popular periodicals after World War II abound with articles and advertisements unanimously heralding nuclear power as the foundation for unlimited economic growth in the United States and throughout the world. More importantly, atomic energy became a metaphor for a centralized, corporatized, capitalistic conception of economic growth which defines bigger as better. Nuclear power fit beautifully within the "progress package," which claimed science and technology as the panacea that would solve any social problem (Gamson and Modigliani 1989). There was little substantive debate about the wonders of the wonders of atomic energy, just about whether such wonders should be used for violent or for peaceful ends.[2] Likewise, there was little discussion about alternative energy supplies, such as solar power, because this fell outside the ideological parameters of the growth package. Nor was there any serious consideration of a "no atom" option. This is an excellent example of third-dimensional power because there were no identifiable social entities overtly or covertly conspiring to limit the debate. Instead, more subtle social currents made certain options more plausible than others.

This progress package remained ideologically powerful as civilian nuclear power developed between 1950 and 1970, and weakened only slightly as the industry began declining in the mid-seventies. Nuclear businesses, government regulators, and the general public all participated (albeit in different ways) in the development and perpetuation of the progress package and its relationship to atomic energy. However, this historical-ideological consensus was no guarantee that civilian nuclear power would quickly become a viable source of electricity. The institutionalization of civilian nuclear power was anything but smooth and steady.

Institutional Issues: Business-Government Relationships

In 1946 Congress passed the Atomic Energy Act, creating the Atomic Energy Commission (AEC) and the Joint Committee on Atomic Energy (JCAE). The JCAE was designed as a congressional watchdog over the AEC, which was in turn charged with promoting and regulating civilian nuclear power in the United States. The five NRC members were appointed by the president, and no more than three could belong to the same political party. It soon became clear that the AEC was just as interested—perhaps more so—in promoting the military rather than the nonmilitary uses of nuclear power.

The Commission itself reflected this contradiction. David Lilienthal, the first AEC chair, wanted to show how military and civilian applications could complement one another (Ford 1982, 35). But commissioner Lewis Strauss believed the AEC should ignore civilian applications and concentrate on military uses. Strauss represented a minority view yet, along with physicist Edward Teller, helped convince President Harry Truman to make the H-bomb a national nuclear priority and an AEC priority. Through the mid-fifties, U.S. energy policy continued giving a higher priority to the destructive capacity of nuclear power rather than to its nondestructive capacity.

AEC proclamations about its commitment to civilian nuclear power were primarily a facade behind which the government could construct plants producing weapons grade plutonium (Clarke 1985). In the late 1940s, only 9 percent of the AEC budget was spent on nonmilitary projects (Loftus 1950). Some observers felt that, if as much time, energy, and money were spent on developing civilian nuclear power, the AEC could have solved the technical problems associated with atomic power plants (Condon 1958). Although this particular claim was probably not true, it accurately portrays the hypocrisy of gestures like the Atoms for Peace and Plowshares programs, which continued to conceal the military focus of nuclear power in the United States. Many in the business community believed that this AEC spotlight on weapons was the single biggest inhibitor of a successful civilian nuclear program (Hodgins 1953).

Regulation and Promotion

The AEC and JCAE slowly became more devoted to the development of civilian nuclear power. In 1954 Congress amended the 1946 Atomic Energy Act to allow increased private ownership of nuclear technology under the supposedly strict oversight and regulation of the AEC. This somewhat eased the technological secrecy the AEC thought necessary because of its role in weapons development (Ford 1982, 41). More importantly, due to statutory prohibitions on direct cash contributions from the government to the private sector, the AEC developed creative avenues for financially subsidizing corporate investment in atomic energy (Dunbar 1958, 276). For example, the AEC labeled many private initiatives as part of its own research so that it could funnel resources (sometimes including cash) to these companies (*Newsweek*, 25 April 1958, 77). It also gave industry access to many of its laboratories at little or no cost (*Business Week*, 18 February 1956, 144). In addition, the government specifically subsidized uranium enrichment studies and other basic research and development by private companies (Hertsgaard 1982). It is unlikely that corporations would have invested this time and money without such subsidies (Commoner 1979).

Perhaps the most important indirect subsidy of the nuclear industry was the AEC's rubber-stamp licensing and regulation of atomic reactors. The statutory language of the 1954 Atomic Energy Act insisted that ensuring public safety was the major goal of AEC regulation, but it became clear that AEC licensing was meant to encourage rather than to inhibit civilian nuclear power. Supporting this, a former AEC attorney said,

> There are about 31 references in the [Atomic Energy Act of 1954] to the "health and safety of the public." But there is not a single reference in the legislative history—the 4,000 pages of reports, testimony and debates before congress relating to this law—as to what these health and safety considerations really were. Nobody really even thought safety was a problem. They assumed that if you just wrote the requirement that it be done properly, it would be done properly. (Ford 1982, 41–42; see also Green and Rosenthal 1963)

Most congressional debate over the 1954 act concerned not public safety but commercial matters, such as who would own certain "discoveries" and how licenses would be granted (Ford 1982, 43). The

AEC basically allowed nuclear utilities (once there were any) to regulate themselves. Reflecting this, the committee chair said that it was senseless for the JCAE to lay down rules now, but that it might in ten to fifteen years if there was a nuclear industry (Ford 1982, 44).

Despite this promise, almost all subsequent AEC/NRC rules lacked any substantive discussion of the technical aspects of nuclear power. The entire Code of Federal Regulations (10 CFR 1–170) deals almost exclusively with the process of licensing a nuclear plant, not with the criteria for a plant worthy of licensing. These technical decisions are left up to the commission itself and, more often, to the three-person Atomic Safety and Licensing Boards (ASLBs) assigned to each application.[3] The ASLBs in turn direct a large staff which does most of the field work and research on a case. The agency also provides an appeals mechanism for ASLB decisions. Only two plants used this rubber-stamp licensing procedure before 1962, but it was a necessary condition for the eventual flourishing of the civilian nuclear industry. This was the foundation for the supportive, nonhostile relationship between the AEC and the industry it was supposed to regulate.

Growing Pains

A cohesive relationship between business and government was necessary for the electrical utility industry to flourish in the early twentieth century (McGuire 1986). This was also true of civilian nuclear power. Between the late fifties and early seventies, the civilian nuclear program took off and established a healthy, if short lived, zenith. This nuclear heyday would have been impossible without the symbiotic relationship between the AEC and the nuclear industry. Anyone who argues that regulatory hostility destroyed civilian atomic energy ignores the fact that there would have been no civilian nuclear energy if not for the gracious support of government regulators. Furthermore, because atomic energy required an enormous economic investment, only large businesses were capable of accepting the AEC's generous offer. This illustrates how the federal government's policies were biased in favor of corporate capitalism. Smaller companies were simply incapable of participating in this government-subsidized opportunity.

The Power Reactor Development Program (PRDP) was the AEC's first attempt to jump-start the civilian nuclear industry.[4] Created under the guidance of David Lillienthal, but advocated more forcefully by Lewis Strauss, the PRDP was a long-term plan to publicly finance private nuclear plants. In its earlier stages, the AEC secretly built small prototype reactors to demonstrate that the technology was commercially viable. When this failed to generate utility interest, the AEC developed a Five-Year Power Program following the 1954 Atomic Energy Act revisions. This program was a more direct subsidy of private reactors, with the AEC providing a majority of the plant's cost. The commission hoped this would encourage many nonsubsidized reactors down the road. The most important reactor built under this program was completed in 1957 at Shippingport, Pennsylvania, the only one of five PRDP projects that would generate electricity primarily for the general public. This 60-megawatt pressurized water reactor was built by Westinghouse for the Duquesne Light Company. The utility spent $5 million for the plant, and the AEC paid the remaining $55 million.

While Shippingport was under construction, three other utilities made plans to build reactors under the PRDP with smaller, though still substantial, cash subsidies from the AEC. Foreshadowing things to come, each of these plants opened five to seven years after its scheduled completion, and two of the projects cost almost twice what was planned.[5] In 1955, on the heels of these three plants, Consolidated Edison of New York and Commonwealth Edison of Illinois announced they would build atomic plants at Indian Point, New York, and Dresden, Illinois, with no monetary contributions from the AEC. The AEC hoped this would be the dawn of a new civilian nuclear era, but that turned out to be wishful thinking. Until the early to mid-sixties, other utilities either would not or could not build reactors without AEC financing. There was still some interest under the PRDP, but not much.

This lack of enthusiasm reflected the continued high cost of nuclear reactors despite all the indirect subsidies provided by the AEC.

The AEC never gave up. In 1959, AEC Chair Glenn Seaborg announced a "ten-year program" that would put the atomic energy industry on the road to financial success. The commission reemphasized this in 1962 when it joined with General Electric and Westinghouse—the largest American nuclear plant manufacturers—to announce that increased federal subsidies of atomic energy would allow it to produce half the country's electricity by the year 2000. There were only four civilian reactors operating at this time, and the AEC hoped these greater subsidies would stimulate utility interest. Far from a hostile relationship, the AEC was acting very much like the executive committee of the nuclear industry (Ford 1982, 76).

In the mid-sixties, two more AEC rule changes greatly assuaged utility fears about nuclear investments. First, the AEC allowed plants to be sited nearer to urban areas, lessening the very expensive transmission from plant to customer. Second, the AEC eliminated any size restriction on atomic reactors. The commission once thought that the industry should practice by operating small plants (less than 400 megawatts). Utilities objected to this because it prevented them from maximizing economies of scale. Thus the AEC started encouraging the construction and operation of untested reactor technologies.

Pubescence

Despite these gyrations by the AEC, utilities continued to waffle on building nuclear plants. Industry representatives made it clear (and the AEC agreed) that civilian reactors should be privately owned, but this philosophical orientation did nothing to generate the capital necessary to build these plants. It was a marketing ploy by GE that really gave the industry a kick in the reactor core. In 1963 GE sold its first "turnkey" reactor to Jersey Central Power and Light. The turnkey program offered utilities completed reactors at a fixed cost guaranteed not to increase with changes in interest rates or construction delays. Utilities only had to pay for the plant and "turn the key"; GE would take care of the designing, building, and financing.

This was a long-term strategy by GE rather than a quick attempt to make profits. Jersey Central paid $66 million for the Oyster Creek turnkey reactor even though it cost GE about $90 million to build (Ford 1982, 63). Westinghouse soon jumped on the turnkey bandwagon and began offering the same instant reactors below cost. In all, the two companies sold 13 turnkey plants between 1963 and 1967. Both corporations may have lost over $1 billion on these turnkey plants or about $80 million for each plant (DelSesto 1979, 90). The AEC acknowledged that the turnkey prices did not cover the manufacturers' costs. For instance, Oyster Creek generated electricity at $161 per kilowatt hour. The nonturnkey Nine Mile Point (New York) reactor, which was similar in size to Oyster Creek and built around the same time, generated electricity at $305 per kilowatt hour (Balogh 1991, 204, 210).

The logic here is similar to that employed by fast-food restaurants that lure customers with discounts on primary items, hoping that they will also purchase nondiscounted side dishes (which carry a higher profit margin) and keep buying them in the future when no discounts are offered. GE and Westinghouse were willing to take large initial losses so that these utilities would continue purchasing nuclear plants even when the discounts were suspended. In fact, during the same 1963–1967 period, they sold twice as many nonturnkey plants. GE and Westinghouse benefited greatly from this gamble, at least in the short run. By 1967, utilities had ordered 75 reactors. Four years later, 716 civilian plants were under way.

Brief Maturity

After 1967, the AEC finally found itself regulating civilian nuclear power instead of just promoting it. Now, it primarily subsidized the industry with its rubber-stamp licensing process. It is very hard to argue convincingly that this process was hostile to the nuclear industry. In the early days of nuclear power, the AEC bent over backwards to make sure private companies could invest profitably in nuclear technology. Later, during what would turn out to be a brief climax for the civilian industry, the AEC/NRC institutionalized a friendly licensing process to make sure the industry flourished.

Building and operating a nuclear plant takes two steps. First, the utility must ask for and receive a construction permit to build the reactor. At this stage the utility must also convince the AEC that the reactor is being built in the best possible location. Once construction is finished, the utility must ask the AEC/NRC for an operating license. No utility has ever been denied an operating license for a completed nuclear plant. There have been frequent delays acquiring these licenses (for instance, at Shoreham and Seabrook) and not all licensed plants operate (like Shoreham), but the AEC/NRC has never seen any reason to deny a license.

Many aspects of the licensing process illustrate the symbiotic relationship between regulator and regulated. Some argue that AEC hearings for permits and licenses were designed merely as a public showcase and were not in any way supposed to slow down plant construction and operation (see Ford 1982, 75). Important decisions were often made by the AEC and the applying utility before formal hearings even began. This predisposition toward supporting nuclear companies was also reflected by the AEC's procedural language. Anyone disputing a plant's construction or operation were labeled intervenors as though they impeded some natural process. Intervenors were systematically denied information by the AEC and utilities, and had numerous procedural constraints placed on them—such as short time periods for filing briefs. Certain events at these hearings could be defined as hostile to corporate interests, but the hostility had nothing to do with the structural arrangement of the AEC/NRC.

Decline

Civilian nuclear power in the United States screeched to a halt in the late seventies. Applications for construction permits, which had peaked in the late sixties, disappeared by 1978. Twelve applications were submitted between 1970 and 1975, nine from 1975 to 1978, and none after that. Similarly, three plants under construction were canceled between 1970 and 1975, and nine plants were canceled from 1975 to 1978 (Ahearne 1987). By 1983 there was as much nuclear capacity delayed in construction as there was capacity on line (Weinberg, et al. 1985, 24). Atomic energy had never become "too cheap to meter," and it accounted for only 10 percent of all electricity generated in the United States (Daneke and Lagassa 1980, 224). Nuclear reactors were also out of service 40 percent of the time — much more than other types of plants (Wasserman 1979, 222), and nuclear power was losing its luster internationally. Despite a strong industry in France (see Campbell 1988; Jasper 1990), a public referendum in Austria derailed the completed Zwentendorf plant, and Sweden announced plans to phase out its dozen reactors by the year 2010 (Weinberg et al. 1985, 224).

In the midst of this decline in the United States, the federal government passed the Energy Reorganization Act of 1974 (42 usc 5841), which abolished the Atomic Energy Commission. The aec was replaced by the Nuclear Regulatory Commission (nrc), which was charged with regulating civilian (and only civilian) atomic power, and the Energy Research and Development Administration (erda), which would promote both nuclear and nonnuclear energy sources. To some supporters of atomic energy, this reorganization epitomized the federal government's regulatory hostility toward utilities. However, there were very few substantive differences between the aec and the nrc, and any noticeable modifications were far from hostile to the industry. The nrc was more or less the same aec wine in a new bottle.

In theory, the act reflected government response to criticisms of the AEC's ability to both promote and regulate nuclear power. In reality, Congress was more and more concerned with the time it took to license a plant, and the "legitimation crisis" within the AEC which might have been exacerbating these delays (Campbell 1988, 51). This crisis was foreshadowed a year earlier when AEC staff members publicly announced that the licensing procedure was just a charade and that most decisions were made before hearings took place (Ebbin and Kasper 1974). Rather than addressing these regulatory contradictions (which it had, after all, helped create), Congress initially tried to eliminate public intervention in all licensing matters. This legislation did not pass, and the Reorganization Act soon followed.

The NRC hoped to reduce levels of intervention by spending more on "safety research." This supposedly showed that it was more conscientious than the AEC. More covertly, the NRC also hoped to discover that requirements on reactor safety were actually too stringent and that construction costs might be significantly lowered (Campbell 1988, 71). Ironically, while trying to assuage growing fears about reactor safety, the NRC simultaneously toughened the criteria by which intervenors would be granted official standing in licensing hearings (Daneke and Lagassa 1980). Unable to eliminate intervention, the NRC urged utilities to be more systematic in reviewing and documenting plant safety. These were not substantive rule changes but merely administrative modifications in the licensing process. Although this may be interpreted as bureaucratic confusion, it falls well short of regulatory hostility.

Three Mile Island

Although it is generally regarded as the turning point for nuclear power in the United States, the 1979 accident at the General Public Utilities plant in Middletown, Pennsylvania, only intensified a decline that had already begun. The Three Mile Island (TMI) accident demonstrated that there was little relationship between the

NRC's rules and the safety of nuclear plants. For example, only weeks before this accident at Three Mile Island, the NRC officially accepted a scientific study proving that reactors were safe and justifying a more streamlined licensing process.[6] The NRC also had just adopted a numeric scale of 1 to 8 to rate the seriousness of potential reactor accidents. The accident at TMI, which had involved a partial meltdown of the reactor core itself, could not be classified. According to this scale, what had happened at TMI was technically impossible.

In the wake of TMI, NRC chair Nunzio Palladino announced an immediate increase in NRC oversight of reactors because the system of voluntary compliance was unreliable.[7] But the NRC's increased oversight and penalties after 1979 did not necessarily translate into better built reactors. It was still more cost-effective for utilities to pay a fine rather than to implement design or construction changes (Temples 1982). The most important change resulting from TMI was a renewed concern with emergency evacuation, but even this was designed more to impress the public than to punish utilities (see Hertsgaard 1982).

This slightly stepped up enforcement did lengthen the licensing process and cost the industry some money, but these trends had begun well before TMI. Construction permits that in the late fifties took about ten months took almost forty months by the late seventies. Similarly, securing an operating license for a completed plant took fifteen months in the late fifties and fifty months in the late seventies (Campbell 1988, 41). Regardless of the relationship between these delays and the TMI accident, nuclear power advocates usually blame the government for killing the nuclear industry with such bureaucratic distractions. This alleged hostility is also frequently cited as the cause of Shoreham's closing and of the many problems at Seabrook—not only demonstrating historical amnesia about the government subsidies of the nuclear industry, but also neglecting specific events at each plant. These events will be addressed in Chapter 3.

Organizational Issues: Corporations and Financing

Even a friendly and supportive regulatory structure did not guarantee the growth of civilian nuclear power in the United States. Nuclear technology was very new and very expensive. This made it difficult for utilities to generate enough capital to invest in atomic plants. In the previous section I demonstrated how government regulators kept enticing private companies to build nuclear plants. These incentives were almost always related to financing. If the government could somehow reduce the financial risk faced by businesses, or help *shift* the financial risk from businesses to other parties, perhaps these organizations would be more likely to take the nuclear plunge.

These sorts of financial obstacles have often been defined as managerial incompetence, both in the industry as a whole and at specific plants such as Shoreham and Seabrook. This rivets any analysis strictly within a corporation while ignoring important relationships between and among different corporations. Better managers, it is argued, would have found a way to get money and build those coveted nuclear plants. However, a more textured analysis recognizes the complex interorganizational landscape in which the nuclear industry grew. Access to investment capital is not just a matter of taking the right MBA course or hitting the ATM. Investment capital is a scarce resource often under the control of a few powerful organizations and not available just for the asking.

The federal government eventually offered enough incentives for profit-minded capitalists to be willing to build nuclear plants. But willingness did not always translate into the ability to build atomic reactors; the money still had to come from somewhere. That "somewhere" is extremely important, although often ignored by chroniclers of America's nuclear power story. Financing details are critical for understanding how the civilian nuclear program unfolded and how this general unfolding intertwines with the specifics of Shoreham and Seabrook.

There was a continual shift in the source of capital for civilian nuclear plants. At first, the government provided most of the capital. After a while, utilities began tapping more of their own internal sources. Later, they turned to external sources such as stock and bond sales. Finally, bank loans became the primary source for atomic capital. Each time the funding source shifted, it created entirely new parameters for government, corporate, and individual decision making. LILCO and PSNH were part of this larger historical change in the flow of capital. Like their counterparts throughout the United States, LILCO and PSNH executives did not have total control over their organizations' actions. Instead, their decisions were very much constrained and leveraged by the changing reality of finance capital. This changing reality was perhaps the greatest contributor to nuclear power's general collapse and to the specific problems encountered at Shoreham and Seabrook.

Direct Government Subsidies

The AEC was well aware of the economic barriers confronting nuclear investment. One AEC staff member referred to atomic power as "a megabuck for a megawatt" (Hodgins 1953, 145). Some early studies commissioned by the AEC showed that there would be little financial benefit from atomic plants even after certain economies of scale kicked in (Loftus 1950). The AEC was quite reluctant to express these opinions, believing it would discourage private investment. Even Robert Oppenheimer had his pessimistic opinion of atomic energy plants "censored" by David Lillienthal (Ford 1982, 33). Although the Atomic Energy Act's 1954 revisions gave nuclear companies easier access to once-secret technologies, it did not confront the purely financial obstacles inhibiting private investment.

The AEC's Power Reactor Development Program tried to address these financial barriers by channeling public money directly to private companies that built civilian reactors. The AEC spent about $20 billion on this program in the 1950s alone (Dunbar 1958, 280). The future nuclear industry had no problem with government intervention in the free market. Newton Steers, president of the Atomic Development Mutual Fund, said, "There isn't a reactor manufacturer in the U.S. who doesn't favor government assistance" (*Time*, 10 February 1958). Still, these abundant federal subsidies generated very little private sector response.

The Price-Anderson Act

One of the most formidable economic hurdles was the exorbitant cost of liability insurance for nuclear plants. Westinghouse vice president Charles Weaver said that these insurance costs by themselves would deter companies from building plants (Ford 1982, 44). Insurance companies estimated that a nuclear accident might require $7 billion in payouts. Insurers were especially worried about enormous third-party payments to people indirectly affected by an accident (Curtis and Hogan 1980, 201, 214). The accuracy of these calculations was impossible to verify, but the mere *belief* in their accuracy was enough to deter insurance coverage. Representatives from the nuclear industry testified before the JCAE that insurance costs were a formidable barrier to the development of nuclear power and that the government might want to help out if it were interested in moving the program ahead (JCAE 1955, 388–389, 403–404).

The JCAE was sensitive to the industry's financial predicament. Representative Melvin Price strongly urged legislation that would help these corporations overcome their fiduciary problems; Senator Clinton Anderson, a former insurance company executive, drafted the actual legislation, which bore both of their names. California Representative Chet Holyfield offered the only noticeable objection. Holyfield wondered why the industry always insisted on controlling civilian power when it was constantly looking for a government handout (JCAE 1955, 69). Holyfield was basically ignored, and the JCAE easily passed the Price-Anderson Act by voice vote. This law limited all liability claims after an accident to $560 million for each civilian atomic power plant. The bulk of this amount ($500 million) would come from the federal government, with the remainder from a pool of private insurers. Any liability claims beyond that $560 million were "illegal." In addition, the law impeded a citizen's right to file negligence suits against any person or organization even remotely connected to an accident (see Ford 1982, 45;).[8]

There was absolutely no rhyme or reason to the $560 million figure. The AEC really wanted unlimited government indemnity for liability claims but eventually proposed a minimum of $200 million, a figure also without any rational basis. Eventually, JCAE staff director James Ramey proposed $500 million because it was between $0 and $1 billion. Senator Anderson accepted this logic, offering basically a $500 million government insurance policy to the nuclear industry for free (JCAE 1955, 120–122). The AEC still would have preferred no ceiling on government coverage but was glad to have the law passed. The nuclear industry and the insurance companies were also pleased (Mazuzan and Walker 1984).

The insurance industry concluded that limiting liability would save utilities $23 million each year in premium costs (Curtis and Hogan 1980, 211). Insurers said nothing about how much money Price-Anderson would save them if there ever were a liability claim. Hertsgaard (1983) and Wood (1983) suggest that this amounted to a $7 billion subsidy of private costs by the government between 1957 and 1970. The Price-Anderson Act is an excellent example of third-dimensional power in which social actors automatically make decisions that serve certain identifiable interests. There was never any doubt that the JCAE would construct the law, the only question was exactly how much public money would be used to subsidize private corporations. Insurance company representatives did appear before the JCAE, insisting that an indemnity bill would be good not just for corporations but for the entire country. But they seemed to be preaching to true believers. Most congresspeople agreed with one industry representative who argued that, because atomic energy was in the best interests of the general public, then the general public should (via the federal government) share in any financial risks (JCAE 1955, 239–241, 252, 527–528).

There was no discussion about who defined this "public interest"; nor did these corporations suggest sharing their profits with the general public. But it was not as if the nuclear industry and insurance companies had to persuade a hostile or even a neutral government that what was best for corporate capitalism (in this case, its nuclear manifestation) was good for America. The JCAE already espoused this "progress package" ideology. Corporations also wielded more overt structural power on the committee. For example, the JCAE chose the indemnity plan over another similar option because the insurance companies preferred it. The committee never examined the strengths or weaknesses of either plan, it merely accepted the position of those who stood to gain most from the new law. Ironically, the Price-Anderson Act had almost no short-term effect on the civilian nuclear industry. Statutes alone could not overcome the still-significant financial impediments to building a nuclear reactor.

Internal Capital

Because the federal government did not want to finance the civilian program indefinitely, the nuclear industry needed other sources of investment capital. Corporations prefer using internally generated capital over external capital because they can maintain better control over its use. Utilities' predominant source of internal capital is electricity rates, which are set and regulated by state government agencies. Electricity rates did not provide an unlimited source of utility revenue, but regulatory agencies had been consistently generous with rate increases since World War II (O'Neill 1959, 2310). Yet even steady rate increases could not always provide the necessary capital for reactors costing upwards of $50 million.

The rate base for a utility is built on (among other things) the company's total operating assets. This base is used as a starting point when regulators consider rate increases. A utility can increase its rate base (hence its income) by building more generating stations, even if these stations provide overcapacity. By including these new plants in the rate base itself, electric companies can accrue profits from new generators far beyond the actual cost of construction. For instance, a 3 percent rate increase would be $1.5 million larger with a base that included a new $50 million plant. Since nuclear plants are very expensive, utilities had good reason to prefer them over conventional power plants—if they could afford them (see Gandara 1977, 82; Campbell 1988, 95–97). However, most regulators did not allow utilities to include construction costs in their rate base until after a plant was finished and producing electricity commercially.[9] Thus the significant construction costs for a reactor would have to be put up front, with excellent paybacks coming years later *if* the plants began operating. The inability to garner reimbursement for construction work in progress caused utilities to rely more on external capital.

Rather than immediately tap expensive external markets, some utilities found other avenues for generating capital. Commonwealth Edison (Illinois) planned to spend $45 million on the Dresden plant, one of the first civilian reactors built without direct AEC subsidy. The utility itself was actually investing only $30 million. The remaining $15 million came from the Nuclear Power Group, composed of seven utilities (including Comm Ed) and Bechtel, a reactor manufacturer. There were two reasons for this strategy. First, Comm Ed did not have $45 million. Second, it was better for the utility's capitalization ratios to have only $30 million on the books; the other $15 million was on the books of the Nuclear Power Group.[10] In a sense, this ad hoc confederation was acting as its own bank so that one of its members could avoid going to a real bank (*Business Week*, 15 September 1956, 188–190).

This creative strategy was neither reproducible elsewhere nor capable of generating the money needed for investment. Comm Ed and Consolidated Edison of New York (builder of Indian Point) were among the largest and wealthiest utilities in the country. Other companies still could not build plants strictly with internal funds. If the civilian program was going to flourish, utilities would have to start tapping into external sources of capital. The stage was set for a shift in financial control from inside utility organizations to outside.

External Capital: From Bonds and Stocks to Bank Loans

With direct government subsidies discontinued and insufficient internal capital, utilities turned to external capital markets to finance their nuclear projects. This was the beginning of the shift in organizational discretion from inside the organization to outside. As utilities became more dependent on external capital, managers found themselves increasingly constrained in directing their companies. This shift in organizational power was slow but constant during the continued expansion of the civilian industry. Managerial decisions were becoming less and less important to the dynamics of nuclear power, while the decisions of rating agencies, institutional investors, and, eventually, banks became more central. As the cost of nuclear reactors increased, so did the power of these external organizations. This changing flow of capital is crucial to understand the rise and fall of civilian nuclear power in the United States, as well as to understand how Shoreham and Seabrook were caught up in these historical currents.

In the postturnkey era, utilities augmented their internal capital with bond and stock sales. As these sales generated a larger percentage of total investment capital, utilities were in a quandary. Because they could not pass on construction costs in progress, the utilities found themselves with increasing debt and stagnant rates. This led to poor debt-to-equity (in the case of bonds) and market-to-book ratios (in the case of stocks). These ratios provide investors with a rough measure of a company's health and the risk of their investment. As these measures became less favorable, utilities had to increase bond interest rates and stock dividends to keep attracting investment. This, of course, further worsened the risk ratios and started another cycle. Rating agencies were involved because they could lower a utility's bond and stock rating at any point in the cycle, just exacerbating the downward financial spiral. For instance, in 1965 only 11 percent of all electrical utilities had bond ratings below investment grade. This number doubled by 1970 and soared to 50 percent by 1975. In 1980, 63 percent of these utilities had ratings below investment grade.[11]

By the early seventies, the bond and stock markets were drying up for nuclear utilities. With 75 percent of all utility investments coming from external capital (up from 58 percent in 1965), the civilian nuclear program was in serious trouble (U.S. Senate 1974, 462). Utilities began canceling orders for nuclear reactors and abandoning projects already begun. This was perfectly understandable because the higher cost and longer construction time of nuclear plants made them vulnerable to the vicious capital cycle outlined above. Not all reactors were canceled or abandoned, but those projects that continued had to find another source of funding. Both construction/ operating costs and licensing delays were increasing.

Bank loans were the only remaining source of capital. The industry had tried for twenty years to avoid bank loans, but now found itself with little choice if it was to continue building nuclear reactors. Some analysts had suggested that electric utilities would flourish financially if they could raise all needed investment capital either internally or though bond and stock sales (Scaff 1960, 525). That was beside the point in the 1970s. Utilities like LILCO and PSNH, which were still determined to build nuclear reactors, turned increasingly to financial organizations for capital. With utilities capital hungry, creditors could set whatever terms and restrictions they wanted on loans. As electric companies relied more and more on bank loans, they transferred their organizational discretion to the financial community.

Banks had a vested interest in lending this money. Many of these loans were enormous, and banks profit from the same economies of scale as other businesses. For instance, banks might prefer making one $10 million loan rather than ten $1 million loans because it is cheaper to administer one loan than ten.[12] Banks were also big investors in the nuclear industry and would not benefit from a total collapse of civilian nuclear power. For example, the Morgan Stanley investment bank became one of the prominent creditors of many nuclear utilities, and it also had a lot of money invested in nuclear manufacturers such as GE and Westinghouse (Hertsgaard 1983, 132). Finally, the increasing delays and cost overruns associated with nuclear reactors might make utilities consistently dependent on more and more loans just to tread water. Creditors benefit most when debtors must keep borrowing capital yet are still able to make all interest and principal payments (see Fitch and Oppenheimer 1970).

Some banks even made the case that it would be in everyone's best interests if utilities acquired most of their capital from banks rather than from other sources. In the early sixties, Edward White, vice president of First National City Bank (later Citibank), argued that state regulators should allow and encourage utilities to consolidate all of their external debt in one place—namely, bank loans. This would cut down on trips to the bond and stock markets, which, he argued, were more "expensive" (albeit not in dollars) to the utilities, and would allow banks "to advise corporate management on major policy decisions" (White 1961, 2619). Electric rate regulations made the financial community nervous because they limited a utility's retained earnings, which slightly increased the risk of loan default. White believed that easing these restrictions would make banks happy, which would make the utilities happy, which would in turn pass the joy on to the general public.

Despite any reservations about regulatory rules, banks began lending enormous amounts of money to electrical utilities in order to finance nuclear plants. Because most of these plants were already under construction, these initial investments established a sort of dependence in which banks stood to lose money if the plants were not completed. Both creditors and borrowers had a vested interest in doing whatever was necessary to license the reactors. However, this mutual interest should not be mistaken for equivalent organizational power. The financial community *wanted* utilities to complete and license reactors, but it would not have disintegrated if the plants had been abandoned. Utilities, on the other hand, *needed* a continual influx of capital in order to finance construction and pay their other bills. Their organizational survival was very much constrained by financiers' willingness to provide capital and by the conditions under which they provided it.

The same sort of vicious cycle established with bond and stock sales soon emerged with bank loans. Common reflectors of organizational strength weakened as utilities increased borrowing. The financial community responded by setting higher interest rates and establishing stricter covenants to compensate for any additional perceived risk. Soon utilities were borrowing money just to pay off other creditors. General interest rates skyrocketed in the 1970s, increasing the leverage of banks over any utilities that still chose to build nuclear plants. These financial issues are often overlooked as primary causes for the rise and fall of civilian nuclear power. The hegemony of the financial community over electrical utilities provides a more textured explanation than simply looking at the internal decisions of utility executives. These general vicissitudes of finance capital probably played the most important role in the specific troubles that would face the Shoreham and Seabrook plants. This is the focus of Chapter 4.

Individual Issues: Public Opposition

Public opposition to nuclear power in the United States germinated very slowly and, even after becoming more significant, focused on particular plants and not on atomic energy in general. American citizens generally accepted the dominant "progress package" ideology of growth and development and marveled at how nuclear power could contribute to the good life. There was an overall trust of the scientific experts who developed nuclear energy and of the utilities that built atomic plants. As people became more skeptical about the wonders of atomic energy, public opposition to nuclear plants began to

grow and play an important role in the decline of the civilian program. In rare cases, such as Shoreham, regular citizens even questioned the correctness of a centralized, corporatized, capitalistic definition of economic growth and development.

It is hard to generalize about an antinuclear movement in the
United States because much of the opposition to nuclear energy was
plant specific. Indeed, the opposition at Seabrook, Shoreham, and a
few other reactors *was* the opposition. Most utilities building atomic
plants faced no significant grassroots opposition, and plants usually
were abandoned for financial reasons. Public opposition should not
be romanticized or treated as the sole factor that killed U.S. nuclear
power. Opposition was crucial to the decline of nuclear power in the
United States, but this decline probably would have occurred anyway, albeit more slowly.

Small Cracks in the Progress Package

Even after the partial meltdown of the Detroit Fermi reactor in 1966,
opposition to nuclear power was insignificant. "More than five weeks
after the accident, the *New York Times* carried a story on what it labeled a 'mishap' at the Fermi reactor. A General Electric official classified what happened as a 'minor perturbation.' No critic of nuclear
power was quoted. . . . Indeed, it would have taken great enterprise
to have found such a critic in 1966 [because] there was no significant
anti-nuclear-power discourse during this era. Nuclear power was, in
general, a nonissue" (Gamson and Modigliani 1989, 14).

Everything remained quiet until the late sixties, when voices
emerged challenging the wonders of atomic energy (Camilleri 1984,
76). This rising skepticism about nuclear power had two sources. The
first was the environmental movement, which grew stronger during
the wave of social protests throughout the sixties and challenged the
ecological damage caused by nuclear power (Camilleri 1984, 82). A
more important source of opposition was the emerging antagonism
of many scientists, including some who were considered historically

supportive of atomic energy (Camilleri 1984, 76). For example, Lawrence Livermore physicists Arthur Tamplin and John Gofman (who had worked on the Manhattan Project) issued a 1969 report claiming that the AEC's acceptable levels of radiation were at least ten times too high (Wasserman 1979, 15). The AEC ignored this report, although it may have had a profound effect on future impressions of nuclear power. These were not just some easily ignored hippies spouting off about ecological suicide, but were nuclear physicists warning about nuclear energy's many problems.

By the mid-1970s, there was an identifiable and growing nuclear opposition in the United States. Some of this opposition focused on the environmental impact of nuclear power, whereas others questioned the corporate accountability of nuclear utilities and the cost-effectiveness of nuclear plants.[13] This growing public opposition did not immediately harm the nuclear industry, but it did contribute to the lengthening of plant licensing hearings, indirectly adding to the already-problematic financial burdens on utilities. However, as I discussed in the last section, these financial predicaments were already under way just as grassroots opposition was beginning to emerge. The opposition sped up rather than caused the decline of civilian nuclear power.

The 1979 accident at Three Mile Island is conventionally identified as the turning point of public opinion on nuclear power. However, as with the modifications in government regulations following this event, it may have been just a catalyst to changes already brewing. Alan Mazur (1988) debunks the impact of TMI by showing that, after a brief swelling of opposition after the accident, public opinion about nuclear power returned to preaccident levels as soon as the media turned off its spotlight. Similarly, the devastating accident at Chernobyl caused only a temporary surge in antinuclear sentiments, which then subsided when the media turned its attention elsewhere. Mazur argues that it was only after people made the conceptual link between nuclear power and nuclear weapons that a sustained nuclear opposition emerged.

Three Mile Island did not have an enormous impact on the grass-roots opposition at Shoreham and Seabrook. Local activism at both plants preceded TMI by many years, and surges in the quantity or quality of actions were independent of the accident. In many ways, Seabrook's opponents were writing the primer for other antinuclear demonstrators in the United States. Shoreham's grassroots opposition, though significant, was not that instrumental in the unprecedented closing of the plant and was eventually rewarded for its "victory" with the highest electric rates in the country. It was actually the totally unprecedented, "official" opposition to Shoreham by the Suffolk County government that dashed LILCO's atomic dreams. I will discuss this further in Chapter 5.

Conclusion

The history of nuclear power influenced Shoreham and Seabrook just as Shoreham and Seabrook contributed to the history of nuclear power. The four main theoretical issues employed to understand these two cases also help explain larger historical patterns. Various manifestations of ideological, institutional, organizational, and individual power all played a part in the slow rise, brief heyday, and rapid decline of civilian nuclear power. Although each of these factors contributed to this story, I believe that the financing questions were the most important to the industry as a whole. Both Shoreham and Seabrook were built at a time when the civilian nuclear program was experiencing sudden and revolutionary changes. Built a few years earlier or later, the scenarios might have been very different. Yet even though these plants occupied roughly the same historical niche, their stories are very different. That is why it is important to move beyond general analyses of nuclear power and compare specific nuclear plants.

CHAPTER THREE

NUCLEAR POWER AND

GOVERNMENT REGULATION

■

In the last chapter I presented the historical bias of government regulation of civilian nuclear power. In this chapter I will illustrate how this bias worked at Shoreham and Seabrook. There is no convincing evidence that regulatory hostility caused the plethora of problems at Shoreham and Seabrook, just as there is no convincing evidence that this hostility helped sabotage the nuclear industry. If anything, federal and state regulators of nuclear reactors and electric utilities were enormously supportive of LILCO, PSNH, Shoreham, and Seabrook. These agencies found creative strategies to help the regulated utilities even when hoisted by their own regulatory petards.

It is not necessary to become entangled in the theoretical debate over how the state serves elite interests to understand this regulatory bias (see Chapter 6). A thorough analysis of Shoreham and Seabrook must be concerned with the first and second dimensions of power (when identifiable individual and organizational actions matter) and with the third dimension of power (where the logic of decision making is more important than the decision makers). I believe these third dimensional elements are the most interesting because they are often overlooked. Government regulation of nuclear power and electrical utilities seems to have an embedded logic that transcends the personal interests of the individuals involved in regulation, although these personal biases are not irrelevant. The regulatory process per-

petuates and legitimates corporate capitalism, often without any noticeable participation by corporations or corporate capitalists.

The Atomic Energy Commission (AEC) and Nuclear Regulatory
Commission (NRC) were the most important federal entities affecting Shoreham and Seabrook. The plants and their owners also interacted with other federal agents, such as the Federal Emergency
Management Agency (FEMA), which was involved in the evacuation
plans for both plants. Shoreham and Seabrook also became enmeshed in the federal court system with the Racketeer-influenced
and Corrupt Organizations (RICO) case filed against LILCO and
with PSNH's bankruptcy. As I mentioned in the first two chapters,
these political arenas are biased toward opening atomic plants and
ensuring the fiduciary health of the plants' owners (roughly defined).[1] Even occasional individual and organizational exceptions to
this rule are usually subdued by the structural logic of government
regulation. Shoreham and Seabrook illustrate this inherent bias.
Both plants were built within a regulatory and judicial system that
worked in their favor. There is no convincing evidence, as some
analyses maintain, that federal regulatory hostility caused the problems at Shoreham and Seabrook. Instead, certain rule changes
within these political arenas (mostly unintended), as well as certain
unprecedented uses of these arenas, challenged the historically
grounded expectations of LILCO and PSNH. Despite this modest
transformation of the political/regulatory framework, the deck remained stacked in favor of the utilities.

AEC/NRC Regulation

The Atomic Energy Commission and the Nuclear Regulatory Commission represent the most significant regulatory bias. Changes in
AEC/NRC rules during the late seventies and early eighties made it
relatively more difficult to construct and license Shoreham and

Seabrook. These changes in no way constituted an *absolute* obstacle to licensing. Indeed, the AEC/NRC went out of their way to guarantee these licenses in spite of their own rules. On one issue after another, the AEC/NRC staff, licensing boards, appeals boards, and the commission contradicted themselves (and each other) and ignored earlier decisions. Shoreham and Seabrook are a perfect window to this structural prejudice, because they went through the process at roughly the same time and with similar salient issues, but interacted with different representatives of the AEC/NRC (except for the commission itself). The parallel decisions rendered in these two independent cases shows that pro-utility bias does not emanate from specific individuals but is embedded in the logic of government regulation.

Shoreham's Construction Permit

Shoreham's construction permit hearings began in September 1970 and concluded in April 1973, after 73 volumes and 15,000 pages of testimony from various proponents and opponents of the plant.[2] LILCO entered these hearings expecting the AEC to employ the same rubber stamp it had wielded for years. LILCO was so sure of this that it spent over $70 million and completed 60 percent of the reactor vessel before it actually got an official building permit (*Newsday*, 21 November 1981, 5). What LILCO had not expected, and failed to forecast, was their opponents' ability to drag out the process for almost three years. Opponents of this nuclear plant worked within AEC rules in ways the AEC never thought possible.

The Lloyd Harbor Study Group (LHSG) was formed in 1969 to challenge LILCO's plan to build a second plant in that wealthy Long Island community.[3] When LILCO scrapped these plans, LHSG continued as the leading, if misnamed, opponent to the Shoreham plant. The "intervenors" probably knew they could not convince the AEC to deny LILCO a construction permit. However, they were determined to delay the process as long as possible in order to expose the licensing procedure as a sham and possibly to increase LILCO's financial pain. LILCO and the AEC constantly berated intervenors for "abusing the system" and relied on procedural mechanisms to suppress substantive debate. For example, one witness was denied standing because his paperwork arrived a few days late (AEC 1970, 249). The licensing board explained that they had to keep the hearings moving along (AEC 1970, 239).

LILCO's vice president for engineering, Matthew Cordaro, said that LHSG's strategy was "like a total fishing expedition," and LILCO lawyer George Freeman remarked that the opponents wanted to delay just for delay's sake (*Newsday*, 15 November 1981, 5). LHSG lawyer Irving Like couldn't have agreed more: "In David and Goliath situations, with a small opposition against a utility, the government, an industry, and ideology, you must resort to [these means]. This is all that's available since the playing field is not level" (*Newsday*, 15 November 1981, 6). Nobel Prize–winning biologist James Watson testified about the unknown hazards of radiation and said that no nuclear plants should be built until they are absolutely safe (*New York Times*, 16 March 1971, 24). A local fisher argued that the thermal pollution from Shoreham would destroy the livelihoods of many small-scale fishers and their families (*New York Times*, 21 March 1971, 62). Dozens and dozens of witnesses appeared, knowing full well they would be ignored, but hoping that these delays would decrease the chances of Shoreham's opening.

Seabrook's Construction Permit

Shoreham's construction permit hearing set an important precedent for the Seabrook hearings, which officially began in 1973 but did not get fully under way until 1975.[4] Members of the Seacoast Anti-Pollution League (SAPL) tried to work within the AEC/NRC rules to delay Seabrook's opening. As with the LHSG, SAPL knew full well it was playing in a rigged game. A second group, the Clamshell Alliance, also participated in the official hearings but complemented this approach with less polite strategies.[5] The ASLB granted Seabrook's construction permit in 1976, suspended the permit temporarily that year (see below), and reinstated it in 1977. Further challenges by intervenors before and after Three Mile Island generated more ASLB hearings, and the permit was again suspended for one month in 1978. Construction continued during these suspensions.

Intervenors focused on PSNH's plan to build huge pipelines, which would bring cool ocean water to the reactor and send heated water back to sea. This design would make unnecessary the huge cooling towers (with billowing steam) often used as a negative icon for nuclear power. Previous opposition in other venues had been quite successful and had forced PSNH to redesign the tunnels in order to minimize environmental damage to coastal wetlands. New Hampshire law allowed communities to veto facilities such as oil refineries, and this law had been used successfully. Henry Bedford (1990, 40–41) believes that Seabrook opponents mistakenly believed that similar victories would be possible before the AEC/NRC and that they failed to realize the more powerful resistance of government regulators, the nuclear industry, and PSNH.

The AEC/NRC was satisfied with PSNH's promise that all environmental concerns would be addressed (see ASLB hearing, 25 August 1975). The commission had never been fond of EPA guidelines on water pollution but had little choice but to enforce them (see Ford 1982). During the hearings, there were strange and sometimes contradictory assessments of the cooling tunnels by some members of the licensing boards and staff. For instance, when an appeals board found some glaring errors in a field report, the staff responded that, if the errors were no larger than a factor of two, it really didn't make a difference. Asked where this "factor of two" rule came from, the staff responded that it had developed the guideline on its own. Again, this has less to do with the idiosyncracies of individuals and more to do with the absence of specific regulatory requirements within the NRC. Before Shoreham and Seabrook, these issues never arose.

Emergency Evacuation

For both Shoreham and Seabrook, the most significant change in the post–Three Mile Island NRC was renewed concern with emergency evacuation in case of a serious plant accident. Although the need for such a plan is loosely spelled out in federal law (see 10 CFR 50.47b), it had always been treated as just another formality. NRC rules say only that a "workable" emergency evacuation plan must be designed by a utility along with its local government, and that review of this plan would take place not at the construction permit hearings but at operating license hearings. This implies that the AEC/NRC never really felt that evacuation would be an important issue, and that no local government would ever fail to cooperate with such a plan. When, during the construction hearings, Shoreham and Seabrook opponents noted the difficulty of evacuating Long Island and the New Hampshire beaches, the AEC ruled its points out of order until the licensing hearings, which were many years away. This demonstrates how powerful organizations set agendas for discussion and eliminate certain positions without making any formal ruling against those positions.

In 1982, Suffolk County (where Shoreham is located) decided to withdraw its support for Shoreham's evacuation plan (see Chapter 5). LILCO responded to this unprecedented act by drawing up its own plan which omitted county participation. This plan would give LILCO representatives certain governmental powers, such as directing traffic and clearing evacuation routes. LILCO argued that the Atomic Energy Act did not give local governments veto power over nuclear plants (LILCO 1984). At first, the NRC was not quite sure what to make of this unique situation. Antinuclear opponents had been warning for years that postponing evacuation plan discussions until after plants were built (and applying for operating licenses) was inviting this sort of trouble. Outspoken NRC Commissioner Victor Gillinsky said that evacuation plans were frequently ignored by licensing boards and the commission itself. "There was an unwillingness to publicize the possibility of accidents because that would lead to a much more frank discussion of accidents," Gillinsky said (*Newsday*, 20 February 1983, 3). At that time, thirty-seven U.S. atomic reactors—including Seabrook—lacked approved evacuation plans.

In the mid-seventies, the congressional Advisory Committee on Reactor Safeguards (ACRS) noted severe misgivings about the successful evacuation of the Seabrook area. Ironically, one of the more vocal ACRS skeptics about successful evacuation was Nunzio Palladino, who would later become a much less skeptical NRC chair. Palladino warned about the proximity of popular local beaches and the limited road network in the area (ACRS transcript, 22 August 1974). But in this pre-TMI period there were few specific rules about emergency evacuation, and PSNH argued that it was responsible only for the "low-population zone" (LPZ) around the area. Of course, because PSNH got to designate the LPZ, it purposely excluded the beaches. The ASLB at Seabrook concurred with PSNH's design and continually rebuffed intervenors who raised this topic (Bedford 1990, 127–129).

Even after TMI, the NRC rarely operated under a consensus about what constituted a proper emergency evacuation plan. Shoreham's licensing and appeals boards were more hesitant to approve LILCO's evacuation plan than the commission itself. Although NRC consultant Thomas Urbanik said that agency would easily approve LILCO's plan (*Newsday*, 22 November 1983, 3), the NRC instead issued a dizzying array of rulings, overturnings, and more rulings. In 1987, the agency finally decided that LILCO's plan was satisfactory, but there was no clear route to this decision.

In February 1985, an ASLB rejected LILCO's emergency evacuation plan. The board cited a recent state supreme court ruling that LILCO could not legally usurp government operations (such as directing traffic) during an evacuation (*Wall Street Journal*, 23 April 1985, 16). The ASLB otherwise felt the plan was feasible (LILCO 1985). In August, another ASLB concurred with this ruling, but criticized the plan itself as inadequate. The full commission seemed to like LILCO's plan (*Newsday*, 28 August 1985, 24).

In October, an appeals board concurred on LILCO's legal inability to assume governmental duties. Most importantly, this panel rejected LILCO's claim that, in an actual emergency, government workers really would participate in evacuation even while saying they wouldn't (LILCO 1985). This introduced the "realism factor" as a possible means to circument governmental noncooperation with evacuation planning, even though the appeals board felt it was inapplicable.

The Cohalan Shuffle and Other Altercations

The idea that governments don't always do what they say crystallized in May 1985, when Suffolk County Executive Peter Cohalan flip-flopped and ordered the county to participate in LILCO's evacuation plan. I will explore more fully Cohalan's shift in the next two chapters, but its immediate effect was to confuse both the supporters and the opponents of Shoreham. Opponents wondered if this abdication was tantamount to surrender because Cohalan had spearheaded the official anti-Shoreham forces (and their unofficial entourage). Shoreham supporters and LILCO read this as a local government imploding from the weight of its unprecedented and possibly illegal actions. In hindsight, both positions were simplistic because they focused too much on the individual actors and not enough on the structural conditions in which these actors performed.

It is unclear how the Cohalan shuffle affected the NRC which, in November 1985, ordered a drill of LILCO's evacuation plan to be overseen by FEMA. The drill would cost about $300,000 and would be paid for by the federal government (*Newsday*, 14 November 1985, 3). This ruling was not unanimous, and the two dissenters expressed their totally unrelated misgivings. James Asseltine felt the drill was a waste of time and money, because the NRC would be able to conclude only that "we think [LILCO] will do the job." But NRC chair Nunzio Palladino objected to the drill because he thought it would further lengthen the overall hearing process. Commissioner Frederick Bernthal said in support of the drill, "It's important to learn if we can evacuate Long Island" (*Newsday*, 15 November 1985, 7).

Although Seabrook's evacuation plan still had the support of most important government officials, the NRC remained incoherent about what specific features made these plans acceptable. This implies that the NRC was never very concerned with plan specifics and promulgated the more stringent evacuation rules as window dressing for an increasingly skeptical public. For example, in October 1981, New Hampshire Governor Hugh Gallen asked NRC chair Nunzio Palladino about the standards the commission would use to evaluate Seabrook's evacuation plan. It took Palladino six months to reply and then only with vague references to "prompt notification" and "time estimates" (Bedford 1990, 135). This did not sound like the same Nunzio Palladino of the ACRS, who seven years earlier had had very specific concerns about Seabrook's evacuation. Once again, the personalities and peculiarities of individual players maybe less important than the organizational and class positions these individuals occupy.

Cooperating governments agreed very little about the specifics of Seabrook's evacuation plan. PSNH was in the enviable position (unlike LILCO) to sit back and let the state work out the details while the utility concentrated on organizational survival (see Chapter 4). Official cooperation in Seabrook's evacuation eliminated some of the legal obstacles raised by LILCO's plan, but unofficial opposition continued to press the ASLB into making decisions without much institutional guidance. Hearings at Seabrook were usually only a few weeks or months behind parallel ones at Shoreham, so there was little time to set administrative precedents on Long Island.

The Seabrook ASLB under Helen Hoyt became a particularly lively forum for debate among the supporters and opponents of emergency evacuation. There was even significant dissent among the supporters, who disagreed about everything from the number of police available to communicating with French-speaking Canadian tourists (see NRC file, "Legal & Adjudicatory Correspondence"). Judge Hoyt seemed particularly irritated with Guy Chichester of the Clamshell

Alliance and Jo Ann Shotwell, assistant attorney general of Massachusetts. Transcripts of the hearings bear a striking resemblance to the rapport between Julius Hoffman and the defendants in the famous Chicago 8 trial (ASLB Hearing, 19 August 1983). Judge Hoyt did not order Mr. Chichester gagged (like Bobby Seale), and the intervenors did not appear in court noticeably stoned, but it was clear that the judge would have preferred that the intervenors remain silent. Seabrook opponents called for Judge Hoyt to step down, but they mistook the third-dimensional biases of the regulatory process for her own idiosyncratic prejudices. The ASLB judges at Shoreham and Helen Hoyt's successor were far less flamboyant, but this had little substantive effect on decisions.

The Realism Factor

After the February 1986 Shoreham drill, an appeals board upheld the February 1985 ASLB decision but thought it should have been more concerned with the substantive shortcomings of the evacuation plan (*Newsday*, 27 March 1986, 7). Directly contradicting this, another appeals board (reviewing the separate August 1985 ASLB decision) *overturned* most of the substantive objections raised about LILCO's plan and ordered yet another ASLB to reexamine these issues (*Newsday*, 23 September 1986, 19). The NRC staff was clearly having problems finding common ground for their rules and regulations. It would be up to the commission itself to work through these organizational contradictions and institute policy changes that allowed the agency to make some kind of decision, regardless of what that decision was.

In February 1987, the full commission announced that the "realism factor," initially advanced by LILCO in 1985, was being considered as official NRC policy. This would allow NRC approval of a utility's emergency evacuation plan, which included government participation even if that government said it wouldn't cooperate, as long as the company made a good faith effort to secure government support (LILCO 1986). Again, the logic was that, in an actual emergency, state and county workers would *realistically* help with evacuation rather than drive away saying, "We told you so." NRC Chief of Operations Victor Stello explicitly said that this proposal had its roots in the Shoreham case (*Newsday*, 25 February 1987, 3). James Asseltine was the only NRC commissioner who objected to the rule change, but his term expired before the final October vote (*New York Times*, 27 February 1987, D4).

Asseltine's dissent shows the importance of individual regulators within biased structure. For example, there was little doubt that President Reagan would not reappoint James Asseltine. He was replaced by Kenneth Rogers, president of the Stevens Institute of Technology (New Jersey) and a board member of a New Jersey utility that operated three atomic plants. Also, in July 1986, Palladino left the NRC and was replaced as chair by Commissioner Lando Zech. Kenneth Carr, another nuclear submarine commander, filled this vacancy. In the final vote, the NRC unanimously adopted the realism factor as policy, thereby amending the 1981 policy concerning emergency evacuation plans. Even without these membership changes, the final decision (if not the vote itself) would have been identical. But that does not mean that a different decision were not possible if, say, there were four other James Asseltines on the NRC.

Summarizing the rule change, new chair Lando Zech said that licensing boards *may* now presume that, during an actual emergency, "state and local governments will indeed act responsibly and take actions to protect the public" (*Newsday*, 30 October 1987, 3). Both Suffolk County and New York State vowed to fight this in court, where, in Governor Cuomo's words, they were good at spotting legal absurdities (*New York Times*, 30 October 1987, A1).[6] This amazing ruling was immediately relevant to Seabrook. The objections of Massachusetts were no longer relevant, and the Seabrook ASLB dealing with evacuation could assume that all local and state governments within the evacuation area would participate in an emergency regardless of their contrasting claims. In fact, this is exactly what happened when Seabrook received its low-power license in 1989. The NRC accepted "the reality that, in an actual emergency, state and local governments will exercise their 'best efforts' to protect the health and safety of the public" (NRC Document #8908030059). This "realism" decision is perhaps the most important confirmation of the structural bias of the federal regulatory process. The NRC clearly neither intended nor expected post-TMI rule changes to give local and state governments this sort of leverage. When governments used these rules to oppose Shoreham and Seabrook, the NRC decided they could be ignored if it was convenient. This does not suggest a regulatory process hostile to the nuclear industry.

Shoreham's Backup Power

One of LILCO's strategies to overcome Shoreham's opposition was to run it at low power (5 percent) before any evacuation plan was approved. LILCO hoped that this low power testing would demonstrate its competence to run Shoreham safely. In addition, once fuel was loaded and a 5 percent power level reached, the plant would be radioactive and many of the advantages of scrapping Shoreham would no longer exist. With the added costs of decommissioning a radioactive plant, LILCO hoped the opposition would see that Shoreham's opening was the least unpleasant alternative. The plant would be radioactive anyway, LILCO would argue, so why not get some electricity from it?

The NRC's structural bias (coupled with organizational confusion) was also apparent in its decisions about this low-power license and the need to have an acceptable backup power supply. As with the rules concerning an emergency evacuation plan, NRC regulations on backup power were unintentionally interfering with its ability to grant LILCO a low-power license. These rules stipulate that, in order to operate a plant beyond 0.001 percent power, off-line backup power units must be available to shut the plant if normal power systems fail. In August 1983, one of LILCO's backup Delaval diesel generators cracked during routine tests. At first the cracks were attributed to poor maintenance (*Newsday*, 15 August 1983, 3), but it was later learned that thirteen other nuclear plants had similar problems with their Delaval units and that the NRC was aware of this but did nothing to "regulate" it (*Newsday*, 16 August 1983, 5). The following week, a second generator (out of three) showed similar cracks (*Newsday*, 23 August 1983, 3). LILCO ordered three new portable generators but planned to fix the Delavals anyway because the replacements would take ten months to arrive (*Newsday*, 6 September 1983, 5).

The NRC responded to these problems with its usual organizational disarray; contradictory responses were developed by a number of licensing boards, appeals boards, and the full commission. In September 1983, a licensing board suggested that Shoreham receive its 5 percent license because the required backup generators were getting fixed—even though NRC rules required them to be in working order during low-power tests. In February 1984, the commission voted 4–1 to approve this 5 percent license (*Newsday*, 13 February 1984, 4). Suffolk and New York State claimed they did not have time to prepare a case. A U.S. district court agreed with them and stayed the NRC's decision (*Newsday*, 26 April 1984, 5).

The county and the state were joined by commissioners James Asseltine and Victor Gillinsky in accusing the NRC of railroading the 5 percent license through. Gillinsky was furious that NRC chair Palladino had been coaching LILCO on getting a license quickly, and Asseltine accused Palladino of assigning licensing boards without the consent of other commissioners (*Newsday*, 16 May 1984, 5). This "undermines the credibility of our licensing hearings and the integrity of our entire regulatory program," Asseltine said (*Wall Street Journal*, 17 May 1984, 15). Palladino replied that as chair it was his mandate not to let "inaction cause any problems at Shoreham" (*Wall Street Journal*, 17 May 1984, 15). Asseltine was not convinced and argued that Palladino should remove himself from all Shoreham hearings because he had shown bias toward the utility.

The absence of regulatory hostility was again apparent in October 1984, when a second licensing board also recommended a 5 percent license for Shoreham, claiming that LILCO's portable diesel generators would do just fine until the permanent replacements arrived (*Newsday*, 30 October 1984, 5). More importantly, the ASLB noted "the unusually heavy economic and financial hardships" facing the utility and agreed with LILCO's assessment that the low power license would help its access to external capital (*Wall Street Journal*, 30 October 1984, 10; see also Chapter 4). In February 1985, the commission voted 4–1 in favor of Shoreham's low power license. In dissent, Asseltine called the vote a ploy to help LILCO raise money and was furious that the NRC would "waive a safety agreement [on backup generators] so that a licensee with financial problems can send a message to Wall Street" (*New York Times*, 13 February 1985, A1).

An appeals board soon "vacated" this ruling after receiving a petition from Suffolk and New York (*New York Times*, 22 February 1985, B1). In June 1985, the full commission denied Shoreham's low power license until the diesel generator problems were solved (*Newsday*, 12 June 1985, 3). No rationale was given for the contradiction between this decision and the one a few months earlier. This became moot, however, when three days later a third licensing board ruled that the diesels' problems had been settled and the patched up Delavals should work adequately for more than a year (even at full power) until the portable replacements — which were very late — arrived (*Newsday*, 26 June 1985, 19). In July 1985, the commission granted Shoreham its low power license (NRC Document #8409050521). The testing began; the plant was radioactive.

Seabrook's Transmission Lines

Environmental Protection Agency (EPA) rules require that a convincing case be made for the environmental advantages (or lack of disadvantages) of siting a nuclear reactor in a specific place. The AEC/NRC is supposed to consider these facts during construction permit hearings, but completely disregarded them at Seabrook (see Stever 1980, 3; Bedford 1990, 49–51). One environmental danger from the reactor was the proposed transmission lines that would link Seabrook to New England's main power grid. These lines could take many routes, some of which crossed environmentally sensitive wetlands. One NRC staff member was assigned specifically to compare the environmental and economic costs of the Seabrook site with a location on the Connecticut shore.

This staff member's strong recommendation for Seabrook was based in part on the cheaper and less ecologically harmful transmission of electricity from New Hampshire to Connecticut than from Connecticut to New Hampshire. It seems that new transmission lines would have to be built to send electricity north (from Connecticut) but not to send it south (from New Hampshire). When questioned about this particular physical principle, the supposed NRC staff expert claimed to actually know very little about New England's electrical grid or how Seabrook tied into it. He apparently knew just as much about gravity. Intervenors moved to exclude his testimony from the official record, but the NRC staff lawyer objected to all the fuss. ASLB chair James Frysiak, never one to side with intervenors, had little choice but to ignore "facts" generated by the staff of his own agency.

These and other objections did not move the NRC into questioning Seabrook's siting, and construction continued while opposition increased. The NRC's actions were a self-fulfilling prophecy because the commission takes into account sunk costs when evaluating a utility's application. In other words, the money a utility spends on a preferred site (even before receiving a permit) is a legitimate criterion for NRC approval of that site. The more a utility spends, the less likely any alternative site will be considered or selected. As PSNH spent more and more on Seabrook, objections to its location became irrelevant regardless of their merits. Ordinary citizens do not have this sort of discretion with, say, the local zoning board. If a person starts building a new deck without official approval, the zoning board will not be swayed by any sunk costs when eventually assessing the application (or when fining the deck builder for code violations).

NRC decisions occasionally stalled and stymied Shoreham and Seabrook's successful construction and licensing, but the larger regulatory framework was always favorable toward these objectives. The NRC's bureaucratic labyrinth was not intended to obstruct nuclear plant licensing, although that was often the result. Some delays were instigated by intervenors using the bureaucracy to their advantage, whereas other delays resulted from an agency that enforced and interpreted rules more stringently than before. Still, the Nuclear Regulatory Commission was not inherently hostile to LILCO, PSNH, Shoreham, and Seabrook. Even after the 1975 Energy Reorganization Act, the NRC continued to be more interested in promoting nuclear power than in regulating nuclear power.

Federal Emergency Management Agency

The Federal Emergency Management Agency (FEMA) is responsible for coordinating U.S. policy within the broad area of "civil defense," which usually concerns nuclear weapons not nuclear power. The connection between these atomic cousins led FEMA to become involved with emergency evacuation planning after the 1981 NRC rule change. FEMA also acted inconsistently toward Shoreham and Seabrook but eventually demonstrated a structural bias favoring the industry. FEMA representatives were far more likely than their NRC counterparts to make fair and impartial judgments concerning Shoreham and Seabrook. But these more neutral displays were often swept up by the biased structural currents of federal regulation.

FEMA and Shoreham

After Suffolk County and New York State declared their opposition to evacuation, FEMA announced that it would not even consider the merits of LILCO's plan because there was no government cooperation (*Newsday*, 18 February 1983, 3). By November 1983, FEMA reversed itself and said that evacuating Shoreham would pose no problems and that it was certain that utility workers and school bus drivers would fulfill their emergency duties rather than assist their own families (*Newsday*, 22 November 1983, 3). FEMA's pendulum swung back again two months later, as it publicly wondered if there was any reason to look at the LILCO plan (*Newsday*, 20 January 1984, 19). Eventually, the agency examined LILCO's plan and found more than fifty substantive deficiencies—excluding the issue of usurping governmental powers—and concluded that no evacuation drill was worthwhile (*Newsday*, 16 March 1984, 3). After pressure from the Department of Energy and LILCO, FEMA finally agreed to conduct a drill early in 1986 (*Newsday*, 23 April 1984, 3; 12 October 1985, 14). FEMA director Samuel Speck admitted that nonparticipation by the state and county would prevent them from drawing definitive conclusions about LILCO's plan.

Nevertheless, on February 13, 1986, LILCO, the NRC, and FEMA tested the company's emergency evacuation plan. The Suffolk legislature unsuccessfully tried to prevent this drill with a series of statutes. The first made it a crime to simulate county officials in an evacuation drill (without authorization); the second proclaimed that the LILCO drill would violate this law (*Newsday*, 8 February 1986, 6). LILCO contended that the drill would not violate this law because utility workers would not be "imitating" county workers; they would only "go to assigned stations to assist [county] officials who won't be there" (*Newsday*, 7 February 1986, 23). U.S. District Court Judge Leonard Wexler barred Suffolk from interfering with the drill in any way, claiming that "it is manifestly clear that Congress by no means intended to allow local governments to frustrate or impede the NRC's ability to evaluate a company's [evacuation plan]." On the day of the drill, an accident "occurred" at 5:40 A.M., and a general emergency was "declared" at 9:30 A.M.—just after the morning rush hour ended. LILCO workers were at their assigned locations "in scenario" but doing nothing. By 4:30 P.M., 118,000 imaginary people had been safely "evacuated" to the Nassau Coliseum forty miles from the plant (*Newsday*, 15 February 1986, 4). LILCO proclaimed the drill a complete success, while Suffolk maintained that it was ridiculous (*Newsday*, 16 February 1986, 6).

After the drill, a debate erupted between Speck and Frank Petrone, who had coordinated FEMA's evaluation of the drill. Petrone wrote in his report that the lack of governmental participation prevented any *definitive* assessment of the plan's ability to protect the public. Speck, who only months before had echoed these sentiments, demanded that Petrone remove that sentence from the final draft. Petrone refused and soon resigned. His objection was deleted from FEMA's final report submitted to the NRC (*Newsday*, 15 April 1986, 5; 20 April 1986, 19; *Wall Street Journal*, 15 April 1986, 64).

Petrone was promptly hired by Suffolk County as a consultant and frequently testified on Suffolk's behalf about the inadequacies of the LILCO evacuation plan and the inconclusiveness of the FEMA drill. FEMA was not enamored of the NRC's realism rule, because it might jeopardize public safety (*New York Times*, 22 March 1987, 1[XXI]). However, once the realism factor was employed, even Petrone's reservations became moot, for they dealt with an "unrealistic" factor. This is another example of second dimensional power, as the NRC managed to remove this substantive issue from the regulatory agenda. The issue simply ceased to exist.

FEMA and Seabrook

Within Seabrook's web of federal regulators, FEMA displayed relatively less bias favoring the utility and the reactor. SAPL members admitted early on that FEMA representatives took seriously the substance of Seabrook's emergency evacuation plan despite PSNH's flippant attitude. In the long run, these impartial decisions had very little impact in the extremely biased regulatory process. FEMA representatives who spoke critically of Seabrook's licensing often found themselves transferred, fired, or discredited (Bedford 1990, 133).

FEMA, the NRC, and New Hampshire tested the Seabrook emergency evacuation plan in February 1986 without participation from Massachusetts and seven local communities. As with the practically simultaneous exercise on Long Island, this drill was designed to maximize success. For instance, the winter date offered fewer complications, because there was no beach traffic and children were all easily found in school. A summertime evacuation would have been far more problematic for these populations. Despite this favorable timing, FEMA presented over two hundred pages of deficiencies to Helen Hoyt's ASLB. Some of the problems included a lack of police officers, failure to notify sea vessels of an emergency, and an inadequate supply of ambulances (Bedford 1990, 152–153).

These reservations disrupted the NRC's ability to grant Seabrook a 5 percent license. When the commission first began discussing the "realism clause" (with regard to Seabrook), FEMA objected that the proposed rule change might jeopardize public safety because it left to pure speculation the adequacy of a utility's evacuation plan. FEMA also worried that such an enormous policy shift might fall beyond the NRC's jurisdiction and necessitate some change in congressional statute.[7] But even with the realism clause, FEMA would have to approve Seabrook's emergency evacuation plan before the NRC could grant a full-power operating license.

Edward Thomas, FEMA's Northeast regional director, was ultimately responsible for approving the evacuation plan. After the initial 1986 drill, Thomas refused to approve the plan despite mounting political pressure. He adamantly maintained that there was no firm evidence that the plan would work and that bathers would have little hope of successfully escaping a Seabrook accident.[8] David McLoughlin, Thomas's immediate boss and a career FEMA bureaucrat, backed him even as general agency support deteriorated (*Boston Globe*, 6 June 1987, 52). Governor Sununu was outraged with Thomas and McLoughlin, and vowed to speak with his pals in the White House about fixing the situation in New Hampshire (*Hampton Union*, 21 November 1986). Perhaps coincidentally (perhaps not),

McLoughlin was replaced in December 1987 by Grant Peterson, whose expertise in FEMA issues seemed less important than his having run Ronald Reagan's Washington state campaign in 1984 (Bedford 1990, 182). Peterson quickly displayed his displeasure at Thomas and made clear that FEMA should not be the lone regulatory holdout over Seabrook's evacuation plan.

Edward Thomas soon found himself transferred to a flood relief position in the southern United States. His replacement, Richard Donovan, witnessed a second and mostly identical evacuation drill in June 1988 and declared it nearly flawless. This drill did not evacuate thousands of bathers at nearby Hampton Beach, and schoolchildren were "simulated" on buses (Bedford 1990, 184). Richard Strome, Governor Sununu's state emergency management director, announced New Hampshire's continuing support for this evacuation plan. In the fall of 1988, FEMA gave the Seabrook ASLB its official approval of the emergency evacuation plan. A few months later, John Sununu was George Bush's chief of staff, Richard Strome assumed the position once held by Edward Thomas, and Wallace Stickney—previously New Hampshire's transportation secretary—became national director of FEMA. Edward Thomas's rogue behavior would probably not be repeated. As with the equally idiosyncratic behavior of NRC commissioners like James Asseltine and Victor Gillinsky, the structural logic of federal regulation, as articulated by individual actors, seems to sweep away all miscreants. Hypothetically, there could have been enough rogue individuals to defy this structural logic, but this never happened in the Shoreham and Seabrook cases.

State Regulation

The New York Public Service Commission (NYPSC) and the New Hampshire Public Utilities Commission (NHPUC) were the primary state government regulators of LILCO, Shoreham, PSNH, and Seabrook. Most utilities are considered quasipublic companies. Such organizations are owned and operated privately and for profit, but they do so with direct government oversight. State regulators must approve rates, construction plans, and other company activities that might affect the "public interest." Theoretically, as with federal regulatory agencies, state regulators are not predisposed toward the interests of the utility (profit) or the interests of the public (low rates, good and safe product); these interests are supposedly balanced. In reality, the NYPSC and NHPUC demonstrated a very noticeable bias toward the regulated utilities and their creditors, and much less concern with the overall public interest. This bias is deeply embedded in the political regulatory process, which operates in the name of "overall economic growth and development" but in reality serves more parochial interests.

LILCO and the NYPSC

The New York PSC basically rubberstamped LILCO's rate increase requests. Between 1970 and 1981, the NYPSC granted LILCO thirteen rate hikes totaling about 72 percent or $521 million. Between 1982 and the one-dollar deal, it approved another 21 percent in increases. The NYPSC also allowed LILCO to divert rate increases requested for specific tasks (such as tree trimming around power lines) to pay for Shoreham. In 1976 LILCO was also permitted to recoup about 5 percent of Shoreham's costs through rate increases (*Newsday*, 20 November 1981, 6). Finally, the NYPSC allowed LILCO to "hide" its investment in the Nine Mile Two plant in upstate New York so that it wouldn't appear as nonproductive capital and jeopardize the utility's credit ratings (*Wall Street Journal*, 30 August 1978, 49).[9]

Although none of this was illegal, it was unprecedented, as was a later NYPSC decision which allowed LILCO to keep its savings from federal tax cuts, normally passed on to customers (*Newsday*, 20 November 1981, 23). Karen Burstein, a NYPSC member from 1978 through 1981, said that the agency often acted like a rubber stamp because it had an enormous workload (*Newsday*, 18 November 1981,

24). Yet even when the NYPSC staff concluded in 1979 that Shoreham was "out of control," the commission still denied a request that LILCO halt construction. It claimed that too much money had already been spent and that stopping the plant would not be good for LILCO (*Newsday*, 18 November 1981, 25; 16 November 1981, 21). This suggests a more endemic philosophy toward corporate health and less a problem of overwork.

The NYPSC also gave LILCO a limited ability to pass on the cost of Shoreham's "construction while in progress" (CWIP). In 1976 the commission allowed up to 5 percent of Shoreham's costs to be included in the rate base. By the end of the decade, LILCO had collected about $255 million through CWIP (McCaffrey 1991, 65). This was an enormous advantage for LILCO. However, the NYPSC did not always make decisions that served LILCO's immediate interests. Some rate increases were approved at less than the sum LILCO requested. In 1979 the NYPSC approved only 3.3 percent of the almost 20 percent increase requested by LILCO and refused to allow another $400 million of CWIP costs into the rate base (*Newsday*, 27 May 1988, 43).

More importantly, the commission ruled in December 1985 that approximately $1.4 billion of Shoreham's $4.2 billion cost was "imprudent" and could never be included in the company's rates. The NYPSC decision cited LILCO mismanagement and lax oversight. It reached these conclusions in part by comparing Shoreham's costs with those of other reactors. It also decided that LILCO could not pass on any costs incurred while the utility halted construction to fix the diesel generators. LILCO replied that comparisons with other nuclear projects were not valid and that many cost overruns and delays were beyond its control (LILCO 1986). The NYPSC did not acknowledge its own lax oversight of LILCO and Shoreham although some agency critics pointed this out during deliberations (*Wall Street Journal*, 14 March 1985, 22). On appeal from LILCO, the State Supreme Court upheld most of the NYPSC's findings but questioned the extent of utility mismanagement. The court believed that factors external to LILCO were extremely important to construction delays, especially those posed by the Suffolk County government.

The NYPSC and the One-Dollar Deal

Shoreham/LILCO opponents used the NYPSC's conclusions to bolster their case. These findings also provided the foundation for the RICO lawsuit filed against LILCO by Suffolk County (see below). But opponents were premature in assuming that the NYPSC's findings reflected a deep antagonism toward LILCO or any dissipation of a rubber-stamp philosophy. This became clear during deliberations over the one-dollar deal to close Shoreham. There was no doubt that the NYPSC would approve the proposed ten-year, 63 percent rate hikes. Only its justification was in question. Not surprisingly, in approving the deal 6–1, the commission said that the deal was the best possible alternative which provided the most financial certainty for LILCO and was best for Long Island's economic growth and development (*Newsday*, 16 September 1988, 3). After being appointed NYPSC chair by Governor Cuomo in December 1987 (Cuomo's fifth appointment out of seven positions), former NRC commissioner Peter Bradford issued a statement that the Shoreham controversy should be settled.

By law, however, the NYPSC cannot approve rate increases for more than three years at a time. This procedural fact contributed to the state legislature's refusal to approve the first one-dollar deal. The second one-dollar deal addressed this legal problem by having LILCO formally request these rate hikes periodically. In March 1989, the utility promptly requested a series of long-term rate increases. Ten days later, the NYPSC approved an immediate 5.4 percent increase followed by 5 percent increases in December 1990 and 1991. After that, pending "official" NYPSC approval, LILCO would receive a 4.5 to 5 percent rate increase each year through 1999 (*Newsday*, 23 March 1989, 5).

The NYPSC's role in approving the one-dollar deal illustrates an embedded, structural bias toward a specific conceptualization of economic growth and development. The New York PSC, like its counterparts elsewhere, was designed first and foremost to perpetuate and legitimize corporate monopoly control of utilities. Regardless of who sat on the commissions at any given time, the logic of utility regulation would work to serve this centralization of electrical power. This systemic logic was not absolute, and at certain identifiable moments the actual members of the commission could have made an impact on regulatory decisions. Given historical proclivities, the NYPSC probably would have made the same decision about the one-dollar deal even if Governor Cuomo had not stacked the commission with personal appointees. But the governor seemed intent on ensuring that the NYPSC support what he thought was a great agreement. The electricity consumers of Long Island did not share the governor's delight.

PSNH and the NHPUC

The same bias was evident in the relationship between the New Hampshire Public Utilities Commission (NHPUC) and PSNH. The NHPUC is also designed primarily to serve the regulated utilities' economic interests. It was never openly hostile to PSNH or Seabrook. Its comprehensive collection of decisions favored the utility and the reactor, despite occasional judgments that unintentionally hampered PSNH. This bias existed regardless of who sat on the commission, but did not preclude more conspicuous biases based on personal views. Again, regulatory bias is not necessarily based on the prejudices of individual regulators—although these prejudices are important.

For example, in 1987 Governor Sununu refused to reappoint Lee Aschliman to the NHPUC. Aschliman had been critical of Seabrook and was at times almost prophetic in foreseeing some of the reactor's troubles, but he was not predisposed against nuclear power or the mandate of utility regulation. Still, John Sununu had no tolerance for Aschliman's attitude and replaced him with Linda Bisson, a former chair of the New Hampshire Business and Industry Association

(BIA) who was instrumental in working out a deal with PSNH to lower electricity rates for the state's largest corporations (see Chapter 5). PSNH allegedly lobbied Sununu to appoint Bisson; with her selection, Sununu had appointed the entire commission (*Boston Globe*, 6 July 1987, 17; 12 July 1987, 41). This probably ensured that the NHPUC would mirror the governor's wishes even though its prior decisions had not given Sununu any legitimate cause for concern.

The NHPUC had given Seabrook its blessing early on and was committed to granting PSNH the ability to complete the project. Rate increase requests were rarely refused or modified, and the commission specifically allowed the utility an above-average rate of return in order to convince the NRC that PSNH was financially qualified to complete Seabrook (Bedford 1990, 8, 97). After granting a large rate increase in 1979, the commission chided some of Seabrook's opponents for causing PSNH's financial problems and reiterated its position that the reactor was a great idea. A few years later, the NHPUC took an official position that finishing the plant and including its construction costs in the rate base would be best for New Hampshire customers (*Boston Globe*, 1 February 1986, 11). It was not clear how the agency reached this interesting conclusion. However, as with the NYPSC and LILCO, the NHPUC's rubber stamp did wear thin as PSNH returned again and again for its "last" rate increase and the utility's explanations became less convincing (Bedford 1990, 110). But this in no way contradicted the NHPUC's historical bias toward the utilities it regulates.

This bias was apparent in the NHPUC's position on CWIP costs. New Hampshire did not legally allow construction costs to be included in a utility's rate base. When PSNH approached the NHPUC about amending this restriction, the agency quickly agreed that a rule change would be best for investors and rate payers because it would make PSNH a more stable company.[10] Assuming the commissioners genuinely believed this, it reflected a belief that what is good for corporate stability is good for everyone. As I mentioned in Chapter 1, this is the dominant ideology in the United States, but it is by no means the only ideology. A solid case could also be made that keeping rates low and decentralizing (and decorporatizing) utilities would be in the best interests of most people in New Hampshire. But regulatory agencies such as the NHPUC would rarely if ever espouse such a philosophy, because it falls outside "normal" ideological parameters. Despite the NHPUC's blessing, the New Hampshire legislature passed an "anti-CWIP" law in 1979 which withstood a series of court challenges by PSNH (*Boston Globe*, 1 September 1987, 42).

Judicial Involvement

The Shoreham and Seabrook stories also involved the judicial system, another government domain. The two most important examples of judicial involvement are the RICO lawsuit filed by Suffolk County against LILCO, and PSNH's bankruptcy. The RICO case ended up in U.S. District Court before Judge Jack Weinstein. PSNH landed in U.S. Bankruptcy Court before Judge James Yacos. I will examine these judicial forums more closely in the remaining chapters, but it is important to introduce them now because they further illustrate this chapter's theme: that government actions are partial to certain definitions of economic growth. In these cases, the courts were no more a neutral referee than any of the other government regulators discussed.

LILCO and RICO — Briefly

In 1986, New York State replaced Suffolk County as the primary official opponent to Shoreham (see Chapter 5). The county soon found itself excluded from all important decisions. In one last gasp to stay involved, Suffolk sued LILCO under the Racketeer-influenced and Corrupt Organizations (RICO) Act for continually lying about Shoreham's completion to the NYPSC in order to obtain rate increases (*Newsday*, 4 March 1987, 7). With treble damages possible under a RICO lawsuit, LILCO could have been liable for about $8.7 billion. Given the December 1985 NYPSC "imprudence" decision, Suffolk stood a fair chance of winning this case. LILCO responded that it might have been mistaken about completion dates, but there was no intention of fraud.

The RICO hearings under Judge Jack Weinstein were filled with twists, turns, and surprises. Verdicts were reached, then overturned by the judge, only to be altered again by a higher court. The timing of the hearings was critical because it actually bridged the two versions of the one-dollar deal that closed Shoreham. There is little doubt that Judge Weinstein much preferred the one-dollar deal to other alternatives for dealing with LILCO and Shoreham, regardless of its fit with legal statutes. Because there were clear winners and losers with the one-dollar deal, the court's actions demonstrated bias toward the interests of the winners. This particular relationship between winners or losers and the law will be discussed in Chapter 6.

PSNH and Bankruptcy—Briefly

When PSNH entered bankruptcy in 1988, most observers felt it would be the epitaph for PSNH, Seabrook, and all those who had poured money into them. This reflected a fundamental misunderstanding of bankruptcy law and the bankruptcy process. Although PSNH as an organization did not survive the bankruptcy, the process clearly favored certain parties at the expense of others. Bankruptcy laws were designed to protect the economic interests of the relatively powerful and overlook the interests of the relatively powerless. Kevin Delaney (1992) argues that bankruptcy is best understood as a political arena in which certain parties have a strategic advantage over others. As with regulatory law, bankruptcy law is concerned only in passing (if that) with the needs of the general public. Instead, what's good for companies, and especially for their creditors, is deemed good for the society as a whole.

Ironically, Seabrook may never have opened if PSNH had not declared bankruptcy, yet bankruptcy was not the first choice of any individuals or organizations with a stake in this case. Creditors did not "force" PSNH into Chapter 11, but neither did the utility pursue it with relish and gusto. Instead, bankruptcy was seen by many parties as the least painful available alternative because bankruptcy legally limits the intrusion of problematic forces, such as official government intervenors (Massachusetts) and grassroots opponents.[11] So, despite the conventional stigma attached to a bankruptcy, it is not always the anathema so feared by organizations, their owners, and their creditors. In the following chapters, I provide details of how the bankruptcy process systemically produces a very predictable cast of winners and losers.

Conclusion

This chapter challenges two popular notions about government regulation of business: first, that regulatory agencies are fair and neutral arbiters of conflict; and second, that any regulatory bias is usually hostile to businesses and capitalism. Neither of these popular notions is supported by the events at Shoreham and Seabrook. With notable and important exceptions, government agents made decisions that perpetuated a centralized, corporate, capitalist definition of economic growth and development and the role of nuclear plants like Shoreham and Seabrook within this growth paradigm. These biases were sometimes obvious (first dimensional), obscure but still identifiable (second dimensional), or completely interwoven with the regulatory foundation (third dimensional).

Structural bias goes beyond the individual actors who compose these government agencies, although the actors themselves are not irrelevant. The NRC/NYPSC need not have been created by corporate capitalists, nor need its members be corporate capitalists, for the agencies to act as advocates of corporate capitalism or to believe that boosting corporate capitalism is the best way to ensure general social prosperity. Even individuals who may be hostile to such an orientation may find themselves swept up in this logic. Federal and state agencies did not display any endemic hostility toward LILCO, PSNH, Shoreham, or Seabrook, although certain specific actions were painful to the utilities and their supporters. Instead, as both cases clearly demonstrate, government regulation of nuclear power and public utilities was designed more to promote the interests of the regulated than to protect the interests of the general citizenry—assuming, of course, that these two groups have different vested interests.

CHAPTER FOUR

NUCLEAR POWER

AND CORPORATE POWER

■

The nuclear industry does not have a product that utilities can afford to buy.
—Tennessee Valley Authority Director David Freeman

Nuclear plants are being built on giant Mastercards.
—Utility analyst Irvin Bupp (December 1983)

The banks basically said take it or leave it.
—Former New Hampshire Yankee President Ed Brown on the "birth" of New
Hampshire Yankee (June 1992)

Another popular explanation for the trials and tribulations of Shoreham and Seabrook is alleged mismanagement by LILCO and PSNH
executives. According to the mismanagement argument, different
decisions by management would have resulted in more organizationally sound outcomes. Presumably, Shoreham would have
opened and PSNH would have stayed alive (with Seabrook opening
more quickly) if only there had been better executive decision making. In this chapter, I challenge this managerial emphasis as overly
simplistic and too concerned with individual actions instead of structural dynamics. In addition, the mismanagement argument places
too much significance on decisions *within* specific organizations
while neglecting decisions *between* and *among* organizations.

The most important interorganizational relationship overlooked
in these cases is the unequal relationship between the utilities and
their creditors. Organizations that control corporate capitalism's col-

lective pursestrings can alter the playing field for other organizations that need large amounts of capital (Glasberg 1989). Nuclear utilities are especially susceptible to the financial community's power because of an atomic reactor's enormous cost. Shoreham and Seabrook were enmeshed in the soaring cost of nuclear energy I discussed in chapter 2. Supposedly *internal* managerial decisions to start building and continue building a nuclear plant are greatly constrained by the ability and willingness of *external* organizations to make loans, how much the loans will cost, and what contingencies accompany the loans.

This flow of capital becomes even more critical when cost estimates prove inaccurate. LILCO and PSNH severely underestimated the cost of building the reactors, which was largely beyond utility control. Whether or not this was due to stupid management misses the more important point: the scarcity of needed capital increased the financial community's influence over Shoreham's and Seabrook's fates. This does not mean that managerial decisions were totally irrelevant or that financiers were omnipotent. But a thorough analysis of Shoreham and Seabrook cannot overlook the strategic power of the financial community in creating and eliminating the available alternatives for LILCO and PSNH.

This relatively simple relationship between borrower and lender is shrouded in layers of complexity. For example, a utility's bond and stock ratings greatly influence its access to capital, yet the rating process itself is highly arbitrary and mysterious. Institutional investment in utilities is also important in the calculus of interorganizational relationships. Furthermore, there is an intricate, intertwined relationship among borrowers, lenders, bond/stock rating agencies, and the regulatory bodies that set consumer rates. Looking strictly at the managers of LILCO and PSNH disregards these important networks and ignores differences in power among organizations. These complex interorganizational relationships are epitomized by the Shoreham deal and PSNH's bankruptcy and demise. I argue that, despite seemingly dissimilar processes and outcomes, both cases show that very similar organizations and individuals emerged as winners and losers.

Securing External Capital

LILCO and PSNH were no different from other utilities in eventually requiring outside capital to build their atomic facilities. The specific mechanisms of obtaining this capital varied greatly during the plants' construction, but each company experienced these changes in roughly the same way. As time elapsed, bond and stock sales became less important, while bank loans grew more important. In addition, obtaining external capital was often highly correlated with obtaining increases in internal capital—primarily through rate increases.

The Early Years: Trying to Avoid Loans

Like most utilities, LILCO had to invest four dollars into its generating and distribution system in order to earn one dollar of revenues (*Newsday*, 22 March 1973, 5). This increased the need for external capital. Building a nuclear plant only exacerbated this appetite for outside financing. Generally, utilities try to use more than 50 percent internal funds for large construction projects; but by 1973, only 40 percent of Shoreham's costs were paid out of internal capital (*Newsday*, 30 October 1977, 3). LILCO borrowed money from banks for much of this early construction, but it more frequently sold bonds or stock in order to raise outside capital. PSNH was a much smaller company than LILCO and was even less capable of internally generating Seabrook's construction costs. In 1974 PSNH projected $600 million of construction costs with only about $77 million cash on hand.[1]

Throughout the seventies, LILCO was unable to match Shoreham's skyrocketing costs with internal funds and was forced to turn more frequently to bond and stock sales. This would lead to a vicious cycle that increased its need for outside capital while making that capital harder to obtain. LILCO's high reliance on external capital led to a decrease in its credit rating; this necessitated that LILCO offer higher interest rates on its bonds, which then tied up an even greater percentage of its revenues in nonconstructive purposes (such as paying bond interest and principal). Raising money through stock sales posed the same problem, because dividend payments were another "nonproductive" use of assets.

PSNH had even less access to bond markets than LILCO. PSNH was a small company and had more trouble than the enormous LILCO obtaining the necessary underwriting for bond and stock sales. The financial community demanded steady rate increases for PSNH and the ability to pass on Seabrook's costs while construction was in progress. Despite these externally generated constraints, and despite New Hampshire's anti-CWIP law (see Chapter 3), PSNH was nevertheless able to raise capital through stock and bond sales. Over the long run, these lower bond ratings contributed significantly to PSNH's long-term debt and influenced its ability to borrow from banks.

The select ratings shown in Table 4–1 illustrate the relative hardships PSNH faced on the bond market. In the enigmatic logic of the ratings agencies, purchasing PSNH bonds was riskier than purchasing LILCO bonds. The slightest change in bond ratings has a significant effect on the cost of borrowed money. For example, a rating

LILCO/PSNH Bond Ratings, 1974–1982

LILCO

5/74	Standard and Poor's (S&P) lowers LILCO's rating from AA to A
6/80	Moody's (M) lowers LILCO's bond rating from A to BAA
7/80	S&P lowers LILCO's bond rating from A− to BBB
5/82	M lowers LILCO's bond rating from BAA to BAA1

Seabrook

2/74	M reduces general bond rating from A to BAA
1/82	S&P lowers general bond rating from BBB− to BB+
6/82	S&P lowers general bond rating from BB+ to BB−
8/82	M lowers general bond rating from BA1 to BA3

Sources: Moody's *Bond Survey* (various issues), Standard & Poor's *Credit Week* (various issues).

lowered from A to BAA means that a company will have to offer an extra 1 percent of interest to offset the increased risk of the bonds. On a $1 billion bond sale, that extra 1 percent translates into an additional $10 million a year in interest payments. As a company ties more of its assets up in interest payments, it risks further lowering its rating, which keeps the cycle going. These financial complications are usually beyond the control of the utilities' managers.

For instance, LILCO managers were well aware in 1973 of the company's addiction to outside capital and tried to make future bond sales more appealing by decreasing its short-term loan debt (*Newsday*, 22 March 1973, 5). However, as the delays at Shoreham increased along with the costs, this debt restructuring became almost impossible. By 1975, LILCO planned to float new thirty-year bonds solely to raise money for retiring old bonds that were coming due (*Wall Street Journal*, 14 May 1975, 25). This meant that Shoreham would be 80 percent externally financed—an unprecedented ratio for utilities (*Wall Street Journal*, 14 May 1975, 25). In 1975 alone, LILCO borrowed $306 million, of which over $40 million went strictly to paying off old debt (*Wall Street Journal*, 24 February 1976, 12). In early 1978, LILCO announced a future stock sale not for raising productive capital, but for refunding prior bond sales (*Wall Street Journal*, 5 March 1978, 31). LILCO management probably preferred none of these choices, but their alternatives had already become greatly constrained by financial conditions.

Inability to Avoid Bank Loans

LILCO and PSNH started relying much more heavily on bank loans to finance construction as the price of Shoreham and Seabrook continued escalating into the 1980s, and the bond/stock markets grew increasingly prohibitive. The utilities did not totally avoid bank loans during the early construction period, but these loans became much more prominent after 1982.[2] For the first time ever, a greater proportion of LILCO's and PSNH's external capital came from bank loans rather than from bond sales (LILCO 1984, 1985, 1986, 1987; PSNH 1983, 1984).

For several reasons companies would prefer raising external capital through stocks and bonds. First, bond interest and stock dividends usually cost less than the interest and other charges of a bank loan. Second, bank loans generally come due faster than bonds. Finally, bank loans are often accompanied by covenants and contingencies which decrease a company's control over its own destiny. More capital-hungry corporations are more susceptible to these stipulations. What frequently appears to be internal managerial incompetence or arrogance might very well be an external organization exercising power.

This financial hegemony came to a head in 1984, a year that proved quite dystopian for both utilities' managerial autonomy. In August, LILCO had no internal funds to satisfy a $90 million bond payment. The company's consortium of lenders agreed to restructure its debt and delay any financial crises until December 1985 (LILCO 1986). This consortium of fourteen banks agreed to lend LILCO $150 million (Chase, Citibank, and Chemical put up about $20 million each) with some outstanding bonds and accounts receivable used as "third mortgage" collateral (LILCO 1984). This refinancing was accompanied by a host of imposed conditions. First, the interest rate on a good chunk of this $150 million would be 2.5 percent greater than the 10.5 percent prime rate, thereby raising the total cost of capital. Second, proceeds from a $100 million junk bond sale (unrelated to this agreement) had to be used to pay off LILCO's $1.2 billion bank debt. Once half of this debt was retired, LILCO could do as it pleased with the cash. Finally, the utility had to suspend its preferred dividend payments at least until December 1985 and continue its March 1984 suspension of common dividend payments (LILCO 1984; *Wall Street Journal*, 24 August 1984, 2).

PSNH actually missed a $5 million May bond payment, and many observers automatically worried out loud about a possible bankruptcy (*Manchester Union Leader*, 4 April 1984, 1). Merrill Lynch then initiated a complex restructuring of PSNH's debt, lining up a consortium of creditors to alleviate the financial crisis. The consortium would provide $35 million in loans and underwrite $90 million of bonds. As with the LILCO refinancing, there was a host of contingencies. First, the bonds would be sold at "junk" levels, at an astronomical 20 percent interest rate. Second, the utility had to suspend its preferred dividends. Third, Seabrook's owners would pay this consortium $150,000 per month in fees on the loan (until it was paid off) and a $2.7 million fee for the bond underwriting. Fourth, utility spending was capped at $3 million per week. Fifth, plans for a second Seabrook reactor were scrapped (*Wall Street Journal*, 23 May 1984, 12; *Boston Globe*, 20 June 1984, 1). Finally, and most importantly, New Hampshire Yankee (NHY) was ostensibly created to assume ownership and control of Seabrook. NHY would be an independent company whose sole mission was building and running the Seabrook nuclear plant. Company ownership was allotted based on ownership of the reactor. NHY would select its own managers, who might or might not come from these utilities. In theory, then, PSNH was no longer the principal owner and builder of Seabrook, although it was the principal owner of NHY.

Managerial Smokescreens

Despite this overt display of external power, most analyses of Shoreham and Seabrook (past and present) continued to focus on alleged internal mismanagement by utility executives. The limits of this explanation became clear in 1984, when both LILCO and PSNH replaced their top executives. Both management changes preceded these crucial refinancings, but neither seemed to affect them very much. In January 1984, most of LILCO's top management resigned. Charles Pierce, board chairperson for most of the Shoreham period and a LILCO employee for thirty-four years, was replaced at the helm by board member William Catacasinos. Catacasinos had no utility management experience, although he was a business consultant to the AEC and a manager at Brookhaven Labs (*Newsday*, 31 January 1984, 4). Catacasinos announced that everything including Shoreham was negotiable except the survival of the company. But LILCO's strategy under his direction was not significantly different than under Pierce's.

In February 1984, PSNH hired William Derrickson as vice president for nuclear engineering. Derrickson was heralded as a brilliant administrator who would make the tough moves necessary to complete Seabrook (*Manchester Union-Leader*, 3 February 1984, 1; 4 February 1984, 1). He quickly fired a thousand employees and established a new emphasis on productivity (Bedford 1990, 120). PSNH and the press (either independently or in unison) referred to Derrickson as a nuclear superstar and a managerial "go-getter" (*Manchester Union-Leader*, 15 April 1984, 1). Derrickson announced that both Seabrook units could be built for under $7.6 billion, much lower than previous estimates.

But the alleged competence of Catacasinos and Derrickson could not alter the interorganizational landscape in which they were "go-getting." Bonds still could not be paid off, and independent consultants continued to predict higher construction costs. Doris Kelley of Merrill Lynch said that LILCO's management change "had nothing to do with one's approach" to the company, and Salomon's Mark Luftig announced that his assessment of the company did not change with the changes at the top (*Newsday*, 1 February 1984, 7). William Derrickson's arrival also failed to impress the banks, which basically said, "Don't ask," when PSNH chief Robert Harrison asked them about refinancing early in 1984 (Bedford 1990, 120).

Managerial saviors Catacasinos and Derrickson did not support the contingencies accompanying the 1984 refinancings because most of them severely compromised managerial autonomy. In Seabrook's case, Robert Hildreth of Merrill Lynch became a very visible and vocal "administrator" of the financial deal. One anonymous Seabrook owner commented that Hildreth "frequently tells us what to do" despite supposedly working for the owners (*Boston Globe*, 3 July 1984,

35). Seabrook's joint owners (especially PSNH) also had no interest in forming New Hampshire Yankee, but Merrill Lynch said there would be no financial deal without this contingency.[3] In January 1985, Hildreth arranged for Drexel, Burnham, Lambert and Kidder Peabody to join Merrill Lynch in handling PSNH's finances. Neither William Derrickson nor anyone else at PSNH had anything to do with this decision, yet it greatly transformed the parameters in which Seabrook was being built (*Boston Globe*, 22 January 1985, 53; 25 January 1985, 17).

Finance Hegemony in Action—Hurricanes and Shuffles

The organizational as well as the physical landscape of Long Island was greatly altered by Hurricane Gloria in September 1985 (see Chapter 5). This catastrophic natural event, clearly beyond the control of LILCO management, beautifully highlights the unequal relationship between lenders and borrowers and demonstrates that seemingly incompetent managerial decisions are often embedded in more complex interorganizational relationships. Gloria left about 81 percent of LILCO's customers without electricity—many for two days, some for two weeks (*Newsday*, 6 October 1985, 4). Those who had lost power were much more likely to bear negative feelings toward the company, many of them for the first time (*Newsday*, 17 November 1985, 3).

Most of this rise in anti-LILCO sentiment focused on managerial incompetence and was not centered on the larger structural constraints in which LILCO management operated. For example, LILCO had made several moves in previous years which exacerbated the problems associated with Gloria. It scaled back tree trimming around power lines, never installed spacer cables and underground lines, and never updated a storm response plan after one thousand workers were laid off in 1984 (*Newsday*, 6 October 1985, 4). In addition to this lack of preparation, many Long Islanders were incensed that LILCO chair William Catacasinos did not race back from his Italian vacation to help clean up.

Most of this alleged mismanagement was intertwined with LILCO's 1984 debt refinancing. Before restructuring, creditors usually demand some sort of signal from the company that it is willing to cut costs wherever possible (*Newsday*, 12 July 1984, 23; 29 July 1984). LILCO's signal to the financial community included its tree-trimming program and about a thousand jobs (not to mention the compensation of those remaining employees). Without this external pressure, LILCO managers might have trimmed more trees, thereby giving Gloria fewer weapons with which to wreak havoc. Similarly, a thousand more workers would have significantly decreased response time to power outages. LILCO could have chosen other signals to impress the financial community, but obviously not all possible signals would be acceptable. Had Gloria veered out to sea rather than bisecting Long Island, the entire Shoreham story may have ended very differently. But I doubt that better managers could have altered the hurricane's path.

Suffolk County Executive Peter Cohalan's May 1985 shuffle (see Chapter 3) also provides a dazzling example of financial hegemony which challenges more one-dimensional accounts of individual and organizational behavior. LILCO refused to pay its property taxes in 1984 after Suffolk County withdrew from Shoreham's evacuation plan. This income shortfall lowered Suffolk's credit rating for the first time in forty-six years. Suffolk needed to sell $24 million in bonds to finance an out-of-control sewer project and now had to offer 8.9 percent interest (instead of 8.6 percent) because of its low rating. That translated into an extra $1 million of interest (*Newsday*, 17 February 1984, 3). LILCO's tax strike also caused Suffolk to borrow $64 million to reimburse local school districts dependent on LILCO's money. Any penalties collected from LILCO for nonpayment would barely cover the interest on this loan (*Newsday*, 12 January 1984, 3).

Cohalan had foreshadowed his shuffle the previous November, after meeting with various investment bankers (*Newsday*, 30 May 1985, 7). Any future reconsideration of Suffolk's opposition, he explained, was simply an effort to mollify (the financial community) before Suffolk's upcoming bond sales; "to let them know we are not crazy, that we are willing to listen to anyone" (*Newsday*, 14 November 1984, 5). Despite this admission, most analyses of Cohalan's shuffle focused on his decisions as an individual rather than on the structural context in which his decisions were made. Cohalan could have maintained his opposition to LILCO's evacuation plan despite Suffolk's credit problems, but his shift is less surprising when the importance of external capital is added to the equation. Here the power of the financial community extends to a supposedly democratic political institution.

The Dialectic of Internal and External Capital

As I discussed in chapter 3, the New York State Public Service Commission and New Hampshire Public Utilities Commission were in no way hostile to the financial needs of LILCO and PSNH. But there was an important connection between the utilities' access to internal capital (rate increases) and their access to external capital (bonds, stocks, loans). Just because the utilities (not to mention Long Island and New Hampshire rate payers) believed electricity prices were adequate did not mean that the financial community believed rates were adequate. The financial community's rating of utility stocks and bonds and its willingness to lend capital were significantly influenced by LILCO's and PSNH's success in increasing rates. Conversely, the willingness of the NYPSC/NHPUC to grant continual rate increases was influenced by the likelihood of completing Shoreham and Seabrook—which was dependent on access to external capital.

LILCO, *the* NYPSC, *and the Financial Community*

Neither LILCO, the financial community, nor the NYPSC tried to obfuscate this relationship between internal and external capital. For instance, LILCO announced that a 1983 interim rate increase request

($91 million) was "the best we could do to satisfy the financial community" (*Wall Street Journal*, 24 March 1983, 46). After LILCO requested a 3.7 percent rate hike in 1985, Shearson Lehman's Ken Crews said that without this increase LILCO would have no access to financial markets (*Newsday*, 14 June 1985, 5). The NYPSC, after granting a $245 million rate increase in 1984, said its "action is a demonstration of regulatory support when the utility is seeking access to financial markets, and sends a positive signal to the financial community" (*Wall Street Journal*, 16 August 1984, 48; *Newsday*, 15 August 1984, 19). The 1984 refinancing of LILCO's debt came just after this rate increase.

Other episodes also illustrate this complex calculus of capital access. LILCO and its creditors amended the 1984 agreement a year later, following another rate hike. This rate increase was accompanied by a NYPSC agreement to begin automatic rate adjustments designed to keep the company financially intact until Shoreham began operating, but not if LILCO went bankrupt. The company believed "that the cash flow resulting from the financial security agreement provides a needed signal to the financial community of continued regulatory commitment to the gradual restoration of the company to sound financial health" (LILCO 1985). The signal was certainly received, and LILCO arranged a new revolving credit agreement with its lending consortium, effectively giving the company until 1988 to pay its debt (LILCO 1986; *Wall Street Journal*, 8 January 1986, 4).

This new agreement was laced with stipulations. First, interest rates on the outstanding debt were raised to 9 percent at a time when general interest rates were falling, and a $2.5 million "fee" was paid to the banks (*Newsday*, 8 January 1986, 3). More importantly, both preferred and common dividend payments would remain suspended until the financial community was satisfied that internal revenues remained consistent with the LILCO/NYPSC financial security adjustment. The bank consortium had to give approval before dividend payments resumed (*Newsday*, 8 January 1986, 3). Certainly LILCO would have been amenable to arranging this refinancing without suspending dividends. However, that alternative simply did not exist as long as the utility was in a subordinate position to the financial community. Similarly, LILCO and the NYPSC might not have wanted to pass on to rate payers Gloria's cleanup costs, but felt there was little choice in order to keep the external capital flowing. In fact, immediately after the storm, Standard & Poor's Douglas Randall said it was unlikely that the financial community would bail out LILCO unless customers paid for a large portion of the cleanup costs (*Wall Street Journal*, 1 October 1985, 8).

Seabrook, the NHPUC, and the Financial Community

Henry Bedford (1990, 97) likens the relationship between internal and external capital to a "financial high-wire act" by PSNH managers. The NHPUC very generously granted rate increases during Seabrook's early construction, when bond and stock sales generated most of PSNH's external capital. But these increases never seemed to totally placate the worries of the financial community, which continually downgraded PSNH stocks and bonds. Even with the optimistic outlook by the NRC, creditors were dismayed at PSNH's inability to pass on CWIP costs and became downright nervous when the state passed the anti-CWIP law in 1979 (Bedford 1990). Lenders made it clear to PSNH that it had to increase its internal capital in order to obtain bond underwriting and loans (Holt 1979, 74). A decade later, just before entering Chapter 11 bankruptcy, PSNH sought a 15 percent rate hike ($70.9 million) "in order to send the proper message to its bondholders who may soon demand payment" on an issuance (*Boston Globe*, 4 October 1987, 79).

As with LILCO, the 1984 refinancing of PSNH's debt was inexorably intertwined with its ability to secure higher rates from the NHPUC. As the deal unfolded, the NHPUC approved a $450 million plan that would satisfy PSNH's (and the financial community's) requirement for about two years (PSNH 1984, 1985).[4] Although PSNH's creditors were far less conspicuous than LILCO's creditors, it seemed clear that there would be no external refinancing without this internal rate increase (*Wall Street Journal*, 14 February 1984, 35; 3 April 1984, 5; 23 May 1984, 12). Robert Hildreth announced that the entire deal might dissolve if Sununu was not reelected that coming November, because the governor seemed eager to appoint utility-friendly people to the NHPUC (*Boston Globe*, 15 September 1984, 20).

Three years later, the NHPUC approved only one-third of PSNH's 14 percent rate increase request. This not-unexpected decision triggered the financial community's stinginess with subsequent debt restructuring. For instance, despite the approved 4.7 percent rate increase, Moody's lowered PSNH's bond rating on the belief that external capital would become even harder to secure (*Boston Globe*, 30 June 1987, 52). This announcement both reflected and caused the banks' future lending decisions. These complex interorganizational dynamics greatly influenced PSNH's strategic decision to enter bankruptcy.

One-Dollar Deals and Bankruptcies—Bad Managers, Good Managers, or Neither?

The inadequacy of the mismanagement approach to Shoreham and Seabrook is highlighted by the one-dollar deal between LILCO and New York State, and by PSNH's bankruptcy. Both of these events can be simultaneously attributed to bad management and to good management. LILCO's failures with Shoreham supposedly reflected bad management, but the one-dollar deal brokered by these "bad managers" gave the company unprecedented organizational vigor. PSNH's managers bungled the Seabrook project and led the utility to bankruptcy shame; but the bankruptcy brought on by these "bad managers" contributed greatly to Seabrook's successful opening. A convincing analysis of Shoreham and Seabrook cannot simply ignore these contradictions. This necessitates moving beyond easily observable individual and organizational behavior and focusing instead on more covert and subtle interorganizational relationships.

The Financial Community and the One-Dollar Deal

According to LILCO, the entire agreement hinged on whether the company's stocks and bonds returned to "investment grade," and on its creditors' willingness to restructure the $3.5 billion in outstanding debt, which included interest payments of about $1 million a day (*Newsday*, 27 May 1988, 5). LILCO said that, if its appeal to the Internal Revenue Service for a huge tax break (because it was, in essence, losing $5.5 billion—minus one dollar—on the deal) fell through, achieving investment-grade ratings would be almost impossible (*Newsday*, 17 June 1988, 5). Similarly, formal approval of the rate hikes by the NYPSC was a necessary condition for Standard & Poor's to increase LILCO's bond ratings (*Newsday*, 21 July 1988, 29).

Company officials and outside analysts generally agreed that the deal's ultimate appeal was to restore a climate of financial certainty and predictability to LILCO's operations; that is, to decrease the company's risk for investors and creditors. LILCO explicitly announced the deal's crucial component was eliminating the endemic financial uncertainty about the company (*Newsday*, 19 June 1988, 4). The deal allowed a $101 million interest payment due in December 1988, to be paid in installments spread over two years (Newsday, 27 May 1988, 5). The company's board of directors, while unanimously approving the agreement, also felt that it was most important to resolve the financial uncertainty (*Newsday*, 16 June 1988, 5).

Outside reviewers echoed this sentiment. Standard & Poor's Edward Graves said, "a negotiated settlement will increase [financial] stability" (*New York Times*, 29 May 1988, 1[XXI]), and Greg Enholm of Salomon Brothers insisted that "the settlement would improve the financial strength of the company" (*Newsday*, 27 May 1988, 39). The key to the deal, according to Al Mazzorano of Dean Witter Reynolds, was "making LILCO a better credit risk" (*Newsday*, 16 June 1988, 5).

Achieving an investment grade credit rating was crucial because it would make external capital (loans and bonds) cheaper. One analyst suggested that such ratings would result in significant savings (*Newsday*, 27 May 1988, 4), and LILCO continually hesitated to change even minor aspects of the original deal (despite increasing opposition) for fear that it would jeopardize these higher ratings (*Newsday*, 17 August 1988, 6). Once again, the discretion of LILCO management was significantly constrained by the complex calculus of capital flows.

It was not surprising that commercial bankers played a significant role in the agreement. Publicly, we saw two hardheaded bargainers—Vincent Tese (representing Governor Cuomo) and William Catacasinos (representing LILCO)—meeting under trying circumstances and manufacturing a fair deal. Only hours before Cuomo's imposed deadline for a deal, Catacasinos visited Tese's Manhattan townhouse, where they rolled up their sleeves and got down to business. Rarely mentioned in these accounts of rugged individualism is that both people had their bankers with them. Merrill Lynch, New York's lead bank throughout negotiations, was represented by Dennis Kelley, whereas Shearson-Lehman (LILCO's lead bank) was represented by Ken Crews (*Newsday*, 29 May 1988, 5).

Between March and May, these banks were represented at every meeting between Catacasinos and Tese, as well as at those between their various associates (*Newsday*, 28 May 1988, 4). It seems safe to say that this agreement was not concocted solely by two independent "managers" who were looking out for their organizations' best interests. Rather, these investment banks played an integral role in shaping and constraining the final accord. Besides protecting their direct financial interests, these banks could suggest strategies that they believed were in the best interests of the entire corporate community. It is hard to prove that the banks actually had veto power over the deal, but the public reports strongly suggest it. The financial community's public endorsement of the agreement demonstrates that their interests were well represented in the negotiations.

A clear signal of the financial community's involvement in shaping this outcome was the almost immediate reconsideration of LILCO's investment status. When the deal was first announced, Salomon Brothers quickly advised investors to "hold" rather than "sell" LILCO stock, and institutional investors began gobbling up 50,000 shares at a time (*Newsday*, 27 May 1988, 39). A few weeks later, Salomon Brothers moved LILCO stock to "buy" status, and Prudential Bache continued recommending the stock as a good buy (*Newsday*, 16 June 1988, 5). Confirming these shifts, Standard & Poor's announced that it would take quick action on raising LILCO's ratings (*Newsday*, 17 June 1988, 5).

The financial community grew more wary as opposition to the agreement grew. Its exuberant optimism was replaced by a more refined analysis of the political-economic landscape. As we have seen throughout the history of Shoreham, decisions by the financial community were often intertwined with the decisions of other organizations, such as the PSC and NRC. By July, Standard & Poor's explicitly said that it would raise LILCO's ratings only if the PSC formally granted the proposed 5 percent rate hikes (*Newsday*, 21 July 1988, 29). Although PSC approval was never in question, this announcement indicated both a commitment to the agreement and a more tentative appraisal of the agreement's overall chances of approval. After the PSC decision to increase electric rates, both Moody's and Standard & Poor's raised LILCO's ratings to just below investment grade — reflecting both increased apprehension and increased support for the goals of the agreement (*Newsday*, 16 September 1988, 3).

The Financial Community and PSNH's Bankruptcy

It is probably a matter of interpretation whether or not the financial community's actions in 1987 "forced" PSNH's bankruptcy. In his book *Strategic Bankruptcy*, Kevin Delaney (1992) examines three celebrated corporate bankruptcies: Johns-Manville, Continental Air, and Texaco. In the Johns-Manville and Texaco cases, the financial community all but forced the companies to declare bankruptcy, whereas no such leverage was applied to Continental. Clearly, there was no manifest directive from the financial community that PSNH declare bankruptcy. But neither was PSNH mimicking Frank Lorenzo's more independent decision to employ Chapter 11 for Continental Air's strategic organizational ends. Rather, decisions by the financial community, highly intertwined with NHPUC actions, severely constrained the options available to PSNH managers. Bankruptcy may not have been the utility's first choice, but it may have been the most attractive alternative available.[5]

It is also important to emphasize that the conventional fear and stigma of bankruptcy may be misplaced and misguided. Bankruptcy can be a very appealing scenario for certain organizations. The popular misconception that bankruptcy is always bad obfuscates those individuals and organizations who can use bankruptcy to their parochial advantage. For instance, local and regional media covering Seabrook began warning about the widespread horrors of bankruptcy before the 1984 refinancing and again before the actual 1988 bankruptcy (see *Manchester Union-Leader*, 4 April 1984, 1; 29 January 1988, 1). The party line was that bankruptcy was such a horrible alternative that creditors, rate payers, and government agencies would do whatever was necessary to avoid it (*Boston Globe*, 31 January 1988, 29; 9 February 1988, 27). PSNH played on this popular perception in order to leverage the NHPUC into granting emergency rate relief and the New Hampshire legislature into revoking the anti-CWIP law.

In reality, bankruptcy was not an unpleasant scenario for many of the organizations involved with PSNH and Seabrook. Dan Scotto of Salomon Brothers said very matter-of-factly that, for big investors and creditors, it did not matter if PSNH's debt restructuring came in or out of bankruptcy (*Boston Globe*, 24 July 1987, 21). Despite publicly denouncing the bankruptcy option in the fall of 1987, PSNH was quietly conversing with bankruptcy specialists in New York about restructuring its debt while under Chapter 11 protection (*Boston Globe*,

7 November 1987, 35). The financial community actually seemed more pleased with PSNH after it declared bankruptcy. By late 1988, the utility's credit rating started inching upward, and its stock price increased over 80 percent. James Bennett, a bankruptcy investment specialist with R. D. Smith, said that PSNH's bankruptcy "is good news for investors" and was urging his clients (mostly institutional) to invest in the company (*Boston Globe*, 3 January 1989, 27).

As the hearings wore on, it became clear which organizations and interests benefited from the bankruptcy and which did not. There were obvious winners and losers both during the process and when the final restructuring plan was approved. In short, loan capital creditors and institutional investors (often the same organizations) benefited the most from the bankruptcy, whereas bondholders, individual investors, and rate payers (in that order) benefited the least. Various specialists and advisers also won big during the three years of bankruptcy. Thus, unlike the classical economic rhetoric, the bankruptcy process is not neutral but biased toward more powerful organizations. These advantages are not necessarily a function of the organizations themselves but are embedded in the structural logic of bankruptcy (see Delaney 1994).

Conflicts of interest between and among parties were evident throughout the hearings. Bigger lenders seemed to be generally favorable toward the process, whereas smaller investors were not (Bankruptcy Document #3017–3437). Battle lines were also drawn between secured creditors (mostly direct lenders) and unsecured creditors (mostly bondholders). Each group had a different set of interests and, more importantly, a different standing in the bankruptcy proceedings. In bankruptcy proceedings, secured creditors have a much higher priority than unsecured creditors and always get paid first. This is why PSNH played the bankruptcy fear card in its last proposal prior to declaring Chapter 11. This July 1987 plan proposed huge concessions from unsecured creditors (*Boston Globe*, 24 July 1987, 21). While unsuccessful, PSNH was hoping that these bondholders would accept the plan in fear of getting something even worse in bankruptcy court.

Throughout the proceedings, Judge Yacos's decisions illustrated the structural bias of bankruptcy. For instance, Yacos ruled early on that rate payers did not have legal standing in the bankruptcy hearings. This was a clear (and rare) victory for unsecured creditors who did not want another party muscling in on their turf at the end of the queue (Bankruptcy Document #518). The New Hampshire Consumer Advocate argued unsuccessfully that rate payers are directly affected by reorganization, especially by its impact on electric rates (Bankruptcy Document #550). Yacos said that bankruptcy law clearly states that local governments should not have veto power over reorganization plans because "there are enormous political pressures on [governments] to force the [lowest possible rates] on creditors and other parties." In a critical ruling, Yacos declared that bankruptcy law takes precedence over state regulatory law in setting the electricity rates of a company in Chapter 11 (*Wall Street Journal*, 25 September 1989, 4). This was a severe setback to New Hampshire consumers.

Yacos established a similar position with those who opposed Seabrook's licensing because of PSNH's unclear ability to pay for decommissioning as NRC law stipulates. The judge maintained that the low-power licensing had nothing to do with the strictly fiduciary relationship between PSNH and its creditors. He was genuinely uninterested in the substantive merits of the argument and focused instead on whether or not the argument should even be considered. Yacos declared that these intervenors "have not taken the steps required under procedural orders [to receive standing]" (Bankruptcy Document #1562). This contradicted an earlier ruling, in which Yacos said, "Even the devil may speak the truth," when denying PSNH's motion to prevent Seabrook opponents from appearing before the court (NRC Document #18810210518).

There were also apparent contradictions in the judge's treatment of services and fees requested by various parties. Some factions had carte blanche approval, whereas others were constantly denied. For example, in the first year alone Yacos allowed PSNH to spend almost $500,000 for legal and financial advisers. The judge also approved over $300,000 in fees for unsecured creditors, about $200,000 for equity holders, and a retainer for the best bankruptcy firm in the country (at $375 per hour) for secured creditors (*Boston Globe*, 2 October 1988, A1). Less than a year into the bankruptcy, the utility had spent about $18 million for advice. This included a minimum $500,000 fee to First Boston Corporation for special advice, despite its already being on retainer by PSNH (Bankruptcy Document #701–3, 743, 1063). Yet Yacos was also in the habit of denying requests he considered frivolous, even though the parties involved did have official standing in the hearings (see Bankruptcy Document #4120–1, 4509).

Again, this is a reflection not on James Yacos's individual personality or predilections, but on the structural logic of bankruptcy. Any judge in his position probably would have rendered similar decisions, serving the same organizational and class interests. Within this more structural framework, seemingly idiosyncratic decisions by Judge Yacos become much more ordinary and almost predictable. James Yacos was not necessarily biased or contradictory, but he represented a process that had built-in bias and contradictions. Yacos did not personally decide on the winners and losers in PSNH's final reorganization. He merely articulated the logic of bankruptcy which favors certain individuals, organizations and ideologies over others.

The Winners: Follow the Money

In both the one-dollar deal and the PSNH bankruptcy, there were clear winners, clear losers, and some who won and lost. Without a doubt, the financial community as represented by lenders and institutional investors won big. Without a doubt, residential and small business rate payers lost big. Individual shareholders fared pretty poorly, whereas unsecured creditors did relatively well. The only real discrepancy concerned the fate of the utilities themselves. LILCO was a definite winner, and PSNH was a definite loser. This final section will deal directly with how the financial community reached the winner's circle in both cases, and indirectly with how LILCO and PSNH met very different fates. The quality of each company's management simply does not figure substantially in this equation.

Winners in the One-Dollar Deal

After the first one-dollar deal fell through, concerned individuals and organizations quickly started forging a replacement package.[6] With Judge Weinstein's RICO rulings filling the air (see Chapters 3 and 6), LILCO and its creditors faced potentially astronomical losses—at least in theory. The second and final one-dollar deal was also satisfactory to the financial community because it was substantively identical to the first deal with only some minor procedural changes. However, given the experience with the original agreement, financiers tempered their optimism a bit more. Greg Enholm of Salomon Brothers would say only, "The deal looks good" (*New York Times*, 1 March 1989, A1).

The institutional investment community was pleased with the deal. It was neither a coincidence nor a preference for powdery snow that sent William Catacasinos on a Colorado ski vacation the day before the second agreement was signed (*Newsday*, 1 March 1989, 40). Instead, Catacasinos had an important meeting with the United Banks of Colorado which, with over 7 million shares, was LILCO's largest institutional investor. Greg Wagenar of First Chicago Investment Advisors, which owned almost 600,000 shares of LILCO common stock, said, "It's a fair deal for the shareholders . . . who must keep a long-term perspective on LILCO stock. I think the rest of the investment community will come around" (*Newsday*, 28 May 1988, 4). Another analyst with one of LILCO's largest institutional investors was also pleased with the settlement: "The deal is fantastic from a shareholder's point of view. Cuomo is coming out smelling like roses and looks like a hero which is fine with me as long as the shareholders get what they deserve" (*Newsday*, 28 May 1988, 4). Exactly which shareholders got what they deserved was made clear by investment broker Donald Smith. "I don't see how it's such a good deal [for shareholders]," he said. "Certainly for most investors, who bought the stock [many years ago], it has been a tremendous disappointment" (*Newsday*, 3 November 1988, 6). By speaking of "most" investors, Smith was referring to people rather than to shares. In terms of this deal with New York State, small LILCO stockholders vociferously opposed the agreement at two different shareholders' meetings in May and November 1988. Their concerns, which revolved around dividend payments rather than share price, did not translate into voting power on the issue.

LILCO's board unanimously and without much discussion approved the deal with New York State (*Newsday*, 15 April 1989, 3). There was some expressed concern with possible small shareholder reaction at their upcoming (June) meeting, but, given past experience, approval of the deal seemed pretty safe. Jack Malvey of Kidder Peabody said he was sure that people would stand up and say the plant should be turned on, but that they didn't carry much weight (*Newsday*, 14 April 1989, 3). Daniel Scotto of Salomon Brothers said that the shareholder vote would be a formality (*Newsday*, 14 April 1989, 4). Both analysts were correct. The June 28 shareholders' meeting was just like the previous ones. The small, individual shareholders voiced practically unanimous objection to closing the plant, and the final vote overwhelmingly supported the agreement.

LILCO's financial health recovered rapidly once the deal was finalized. In April, for example, Shearson-Lehman bought $375 million of four-year debentures at an 11 percent interest rate. This was the first time LILCO had successfully floated bonds in almost three years. The successful sale eased certain managerial constraints by reducing LILCO's addiction to more expensive loan capital. This debenture sale also gave LILCO access to $88 million in reduced rate loans within ninety days—with access to $200 million more sometime thereafter. LILCO obtained these loans in January, *under the condition* that it first generate some capital through bond sales (*Newsday*, 11 April 1989, 25).

Toward the end of June, LILCO's bank consortium (led by Chase Manhattan, Citibank, and Chemical), restructured $325 million in pending loan payments, including a lowering of interest rates on certain segments of the principal. In addition, the utility received an extended $300 million credit line (*Newsday*, 21 June 1989, 25). Despite ratings still just shy of investment grade, institutional investors began rapidly buying the 11 percent debentures (*Newsday*, 30 June 1989, 36). On July 25, LILCO announced that it was now ready to pay $390 million in back-preferred dividends and would resume normal preferred payments on September 1 (*Newsday*, 26 July 1989, 7). LILCO resumed common dividends of $.25 per share on October 1—the first time such payments had been made in five and a half years.

This organizational recovery is especially astonishing because most conventional perspectives often compared LILCO executives to The Three Stooges. How could such bunglers miraculously invigorate this supposedly moribund company? It is impossible to answer this question by focusing on a company's managers. Instead, the one-dollar deal was produced by a complex dynamic among LILCO, the financial community, and other organizations. For instance, the earlier restructuring deals stipulated that LILCO's banks had to approve any resumption of stock dividends. To say, as the Cuomo administration did, that dividend resumption was an example of management arrogance (*Newsday*, 17 March 1989, 25) obfuscates the complex structural relationship between LILCO and its creditors.

Winners in PSNH's Demise

Although the circumstances were very different, the same winners and losers emerged from PSNH's bankruptcy except for PSNH, which "died" on May 17, 1991, despite being the titular owner of the fully licensed and functioning Seabrook reactor. The bankruptcy ended when Northeast Utilities (NU) purchased PSNH's assets (including its share of Seabrook) for $2.3 billion and instituted seven years of mandatory rate hikes totaling about 43 percent.[7] Interestingly,

NU's offer was not the best, on paper anyway, for either PSNH's owners or its customers. By October 1989, four different reorganization
plans were offered to PSNH's creditors. PSNH's own plan called for
no change in ownership and an average 4.8 percent rate hike each
year for seven years. NU offered $2.25 billion for PSNH's assets and
seven years of rate hikes at 5.5 percent per annum. New England
Electrical Systems offered $2 billion and 3.8 percent rate hikes per
year for seven years. United Illuminating (UI) proposed $2.29 billion
to purchase PSNH and 5.5 percent rate hikes each year for five years.
None of the plans seemed to satisfy everyone, although the biggest discrepancy seemed to be the plans' impact on rate payers, who weren't
even party to the hearings (*Boston Globe*, 25 October 1989, 75).

NU emerged as the frontrunner even before raising its offer to $2.3
billion. By late 1989, all classes of creditors preferred NU's offer despite the possibility of a higher bid from UI, and PSNH eventually accepted NU's proposed takeover. The state of New Hampshire threw
its relatively insignificant backing to NU's plan despite higher electricity rates than under all other proposals. The only objections to this
plan came from rate payers. The final decision was now with the
court: "Even though all parties [in this case] have now joined in supporting the settlement . . . the court still has to independently [de-
termine] whether the settlement is appropriate . . . [and] if it meets
with the intent of congress in establishing Chapter 11. Courts have
sometimes eliminated obstacles to successful reorganization when
appropriate, especially when there is some agreed on settlement"
(Bankruptcy Document #1129; NRC Document #8905040135).

This emerging consensus among bankruptcy parties seemed to
have less to do with NU as a utility and more to do with its primary
benefactor, Citicorp. Citicorp had relatively little financial stake in
either Seabrook or PSNH but chose to help NU acquire PSNH.[8] By
purchasing PSNH, NU hoped to control 65 to 90 percent (depending
on demand) of electrical transmission in New England (*Boston
Globe*, 17 May 1991, 63). Of course, NU could not just pull $2.3 bil-
lion out of its pockets. Its ability to tender an offer for PSNH became

possible only with Citicorp's willingness to provide financial back-ing. The quick acceptance of NU's plan by PSNH's large creditors and institutional stockholders was not surprising, because they stood to make out handsomely with this offer. Citicorp estimated that se-cured and unsecured creditors would recoup between 90 to 100 per-cent of their investment, while shareholders would receive about 25 percent (*Boston Globe*, 28 July 1989, 21; Bankruptcy Document #2340).[9] This estimate did not distinguish between individual and in-stitutional shareholders who purchase stock for very different rea-sons. The Citicorp estimate focused on withheld dividends, which are important only to individual investors. Institutional investors are more interested in stock prices, which continually increased during PSNH's bankruptcy. As with the Shoreham deal, institutional share-holders, many of them large financial organizations, fared quite well under PSNH's reorganization.

Large creditors also exercised subtle influence over New Hamp-shire's official position on the various reorganization plans. The state clearly supported the NU offer and noted that the financial commu-nity's preference was paramount in its position (Bankruptcy Docu-ment #2997, 2879, 2900; *Boston Globe*, 17 December 1989, NH2). There is no evidence, unlike the Shoreham deal, that creditors "told" or "coerced" New Hampshire into taking this position. Instead, New Hampshire's position illustrates third-dimensional power. The state defined its interests (and its citizens' interests) as congruent with the interests of the financial community, without any overt influence or pressure by the financial community.

Conclusion

Blaming LILCO and PSNH managers for the trials and tribulations of Shoreham and Seabrook neglects the larger structural framework encompassing these organizations and their leaders. Surely both companies' managers made ill-advised decisions throughout the twenty-five-year histories of the plants. But internal managerial decisions alone did not cause these problems, and they were often made under highly constrained conditions. The final outcomes at Shoreham and Seabrook belie a simplistic managerial analysis. LILCO and its supposedly bad managers emerged as a very healthy company, and PSNH and its supposedly bad managers got Seabrook on line.

A strictly managerial analysis overlooks the unequal relationship between utilities and the financial community. The financial community is by no means omniscient or omnipotent. Many of the predictions it made about Shoreham/LILCO and PSNH/Seabrook were incorrect. For instance, many banks considered Mario Cuomo the biggest obstacle to a successful outcome at Shoreham, such that his public rhetoric often had an instant effect on bond and credit ratings. But Governor Cuomo turned into the leading advocate of the one-dollar deal, which clearly met with the banks' approval. Whether or not the financial community's crystal ball correctly reads the future has no bearing on its ability to control corporate capitalism's collective pursestrings and to constrain organizations that want to dip into the till. Power is related not necessarily to truth or fairness but to control of strategic resources.

LILCO and PSNH were both capital hungry companies. Every time they secured external capital they lost more organizational autonomy. Creditors did not step in and control LILCO and PSNH, but they did construct and eliminate the alternatives available to the utilities. This vulnerability to creditors might have been reduced if LILCO and PSNH had chosen to stop building Shoreham and Seabrook. But even this decision became a victim of financial hegemony once the banks had invested significant funds in the projects. The banks wanted Shoreham and Seabrook to open because the utilities had either limited or no legal ability to pass on construction costs to rate payers before the plants were operational. LILCO and PSNH would not be able to repay the banks without these rate increases.

The financial community was central (albeit quiet) in developing and orchestrating the one-dollar deal to close Shoreham. It had the power to bend and shape the political-economic landscape to serve its own interests and the interests of most large investors. Long Island rate payers (primarily residences and small businesses) had no say in this deal. In New Hampshire, the financial community's role in PSNH's bankruptcy was far less central, although it did eliminate many of the alternatives available to the company's management. Contrary to conventional wisdom, PSNH's large creditors and institutional investors did not fear bankruptcy, because Chapter 11 laws were designed to protect their interests. Smaller creditors and individual owners were more wary of the bankruptcy option, and rate payers were not even invited to the party. In both of these very different scenarios, the same winners and losers emerged. Powerful organizations like large banks win by shifting their financial risk onto less powerful factions. This can occur in a format like the one-dollar deal, in which the banks are directly involved, or in a process like bankruptcy, which is structurally designed to serve the banks' interests even without their direct involvement. Rate payers and other regular folks lose because they are either quietly denied access and ignored (the one-dollar deal) or legally excluded from participating (Chapter 11).

With these patterns of winners and losers in mind, the popular emphasis on managerial decision making becomes dangerous instead of just inaccurate. If it is just bad executive decisions that caused these biased outcomes, then all that is necessary to prevent future disasters (from the nonpowerful's point of view) is to replace bad managers with good managers. Then everybody will be happy because the "soulful corporation" will respond to social needs rather than to the parochial needs of capital. But if instead there is an embedded logic that regularly (but not exclusively) serves parochial interests, then avoiding future disasters (from the nonpowerful's point of view) may require a serious overhaul of the entire political economy. This is not the way most of us are used to thinking about social phenomena.

CHAPTER FIVE

NUCLEAR POWER AND LOCAL

POLITICAL ECONOMIES

■

The Shoreham plant was torn down because "the people willed it."
—U.S. District Court Judge Jack Weinstein (*Newsday*, 23 March 1989)

Seabrook's support was "a slight majority but the right majority."
—New Hampshire Yankee President Ted Feigenbaum (June 1992)

In Chapters 2 through 4 I illustrated the similarities between Shoreham and Seabrook while challenging some of the conventional explanations for these troubled nuclear plants. In this chapter, I critique another standard theory for why Shoreham and Seabrook faced
so many problems: the efficacy of grassroots opposition, or "people
power." Here we will finally explore the key differences between
Shoreham and Seabrook and why they met such different fates. I will
propose ideas rarely considered in the dialogue about American civilian nuclear power. As a result, it will be the longest and most intricate part of the book. This chapter also more prominently reflects the
book's empirical bias toward Shoreham, because this case generated
the most interesting and unprecedented events related to "people
power."

There was significant grassroots opposition to Shoreham and
Seabrook from individuals and groups living near the plants. This
opposition was sometimes idiosyncratic, but more frequently it was
organized and orchestrated. This "unofficial" opposition (as op-

posed to the "official" opposition from governments) involved a patchwork of privately funded individuals and organizations with very different motivations for opposing the reactors. There was an unusually high quantity and quality of unofficial opposition at Shoreham and Seabrook but this significant grassroots opposition (like mismanagement and regulatory changes) was not a *sufficient* explanation for the problems at Seabrook and the unprecedented events at Shoreham.

The efficacy of grassroots opposition to Shoreham and Seabrook was dependent on the local political-economic context of each plant. There are three general components of these local political economies. First, how important and effective was unofficial grass-roots opposition to each plant and the companies that built them? Second, how important and effective was official opposition of local and state political elites to these plants and the utilities that built them? Third, what was the relationship between opposition to the plants (both unofficial and official) and the local growth coalitions and ideologies on Long Island and in southern New Hampshire?

These are empirical questions which can be answered in many different ways depending on the circumstances. The theoretical importance of these questions is asking them in the first place and considering how they interact. For example, official opposition to a nuclear plant is qualitatively different from unofficial opposition. In addition, the existence of official opposition by local governments may be unrelated to public opinion or the personal characteristics of local elites. Instead, opposition may emerge from the particular structural and ideological characteristics of local growth paradigms and coalitions. Finally, the apparent power of both unofficial and official opposition to a plant may depend on the relative strength of the local growth coalition. This chapter attempts to put "people power" in its proper perspective by looking at the interrelationships among grassroots movements, official opposition, and local growth coalitions in explaining why Shoreham closed and Seabrook opened.

Nuclear Power and People Power

As I discussed in Chapter 2, grassroots opposition often receives excessive credit for the decline of American civilian nuclear power. Unofficial nuclear challenges certainly contributed to the decline but they did not cause it. Increasing material costs, skyrocketing interest rates, waning electricity demand, and bureaucratic confusion all took their toll on the industry, irrespective of grassroots pressure. Many plants that were canceled or abandoned since 1975 had little or no grassroots opposition, and many closings also preceded the Three Mile Island accident.

Comparing Shoreham's and Seabrook's grassroots opposition also challenges the importance of people power, especially its role in closing Shoreham. In its most romanticized version, Long Island small business owners, homemakers, environmentalists, and intellectuals decided they did not want a nuclear reactor nearby and formed a complex network of alliances which fought for their goal despite formidable resistance from more powerful social actors. However, this story ignores the staggering defeats for these same people due to the one-dollar deal. If people power alone was that important, then Seabrook also should have closed. Grassroots adversaries in New Hampshire and Massachusetts were certainly equal to those on Long Island and were in some respects more effective in educating, organizing, and agitating against Seabrook. And many plants with a great deal of grassroots opposition—such as Diablo Canyon, California, and Limerick, Pennsylvania—also opened. Other plants with no opposition closed. The existence and apparent strength of grassroots opposition does not by itself predict whether or not a nuclear reactor will open.

Shoreham: The Early Years

Along with its plan to build Shoreham, LILCO hoped to build another reactor at Lloyd Harbor, also on the Long Island Sound but only thirty miles from New York City. In retrospect, this turned out to be a crucial error. The residents of wealthy, elite Lloyd Harbor had no desire for a nuclear plant cluttering up their neighborhood, and, more importantly, they had the resources to fight it. Lloyd Harbor mayor George Barclay said the plant would offer massive tax relief and spur economic development (*New York Times*, 10 October 1967, 31). But the residents of Lloyd Harbor needed no tax relief and preferred to invest in economic growth and development rather than live near it.

Some of these residents formed the Lloyd Harbor Study Group (LHSG) which I discussed briefly in Chapter 3. By 1969 the group had developed an impressive list of nuclear opponents, including a Grumman executive who said that, because insurance rates for this plant would be almost ten times the industry average, it was probably unsafe. Scientists from the Cold Spring Harbor Laboratories and from the State University of New York at Stony Brook argued that little was known about the effects of radiation on everything from algae to zoospores (*New York Times*, 9 November 1969, 27). Before beginning the construction hearings on Shoreham, LILCO quietly withdrew its Lloyd Harbor plans and hoped these people would go back to a more sedentary lifestyle (*Newsday*, 16 November 1981, 21).

But the LHSG did not die when the Lloyd Harbor reactor was scrapped. Rather, it became the focal point for almost all anti-Shoreham activism in the 1970s, including the regulatory opposition examined in Chapter 3. It was the LHSG that obtained official intervenor status before the AEC and argued against granting LILCO a construction permit for Shoreham. Although opposition was originally rooted in a "not in my backyard" (NIMBY) perspective, many LHSG members became sensitized to larger issues concerning civilian nuclear power. It is quite likely that there would have been significantly less organized opposition to Shoreham (in these early days) had LILCO not planned the Lloyd Harbor reactor.

Working-class, blue-collar Shoreham turned out to be much less hostile to its future reactor, and many locals were incensed that rich "city bums" from Lloyd Harbor were making a big issue out of the plant and that "it's no wonder the indians fought against [their ancestors]" (*New York Times*, 1 October 1970, 43). The village's chamber of commerce predicted that the local tax rate would drop from $30 to $6 per $100 of assessed value if the plant operated. Local schools would also reap a windfall from the reactor's property taxes.[1] Overall public opinion about the plant was at best mixed through the mid-seventies (*Newsday*, 16 November 1981, 22). As explained in Chapter 3, the Lloyd Harbor Study Group was very active during LILCO's construction permit hearings before the AEC. But organized opposition waned after the construction permit was granted in 1973. There was a brief swell of public opposition immediately after Three Mile Island, but this quickly returned to earlier levels, as did similar swells throughout the country (see Chapter 2). Right after Three Mile Island (April 1979), 27 percent of Long Islanders favored completing Shoreham, while 68 percent opposed it. In June, this had changed to 30 percent versus 61 percent and to 42 percent versus 55 percent by July (*Newsday*, 23 July 1979, 5).

Three years later there was a decrease in both those who supported (36 percent) and those who opposed (46 percent) the plant (*Newsday*, 16 July 1982, 31). People's indecision was growing. However, there was little organized grassroots opposition to Shoreham or LILCO at this time. Even after 1982, when the Suffolk County government launched its unprecedented opposition to the Shoreham emergency evacuation plan, there was very little increase in anti-Shoreham public opinion or organized grassroots opposition (although the Lloyd Harbor Study Group had been replaced by the Shoreham Opponents Coalition). Overall public opinion following Suffolk's opposition remained relatively constant at 35 percent favoring Shoreham, 55 percent opposing it, and 10 percent undecided. This level of grassroots opposition to a nuclear power plant was practically unheard of in the late seventies and early eighties. Not including Seabrook.

Seabrook: The Early Years[2]

The Seacoast Anti-Pollution League (SAPL) was both similar to and different from the Lloyd Harbor Study Group. Founded in the late sixties, SAPL was an extremely small group of highly educated middle-class people who opposed Seabrook and planned to work within the legal-regulatory system. But, SAPL was poorer than the early LHSG and not originally grounded in a NIMBY philosophy. Seabrook's unofficial opposition was more concerned with protecting New Hampshire's small coastal ecosystem from overdevelopment. After Seabrook received its construction permit in 1976, there was a significant split in organized grassroots opposition to the plant. Most opponents wanted to continue the legal-regulatory strategy of the past, not unlike what Shoreham opponents chose on Long Island. However, there was a growing surge of opponents who preferred more direct, confrontational tactics to educate and organize New Englanders. Out of this rift, the Clamshell Alliance emerged and almost immediately staged a six-hundred-person "takeover" of the construction site, leading to the arrest of eighteen. The titular leader of the Clamshell Alliance, Guy Chichester, had irritated many Seabrook opponents during his brief reign as SAPL president with his fondness for confrontation and civil disobedience.

Thus Seabrook's grassroots opposition used a double-barreled approach. One barrel was the polite, intellectual, increasingly innovative SAPL, which quietly worked through the system, hoping that the force of reason would eventually sway official decisions. The other barrel was an impolite, charismatic, increasingly innovative Clamshell Alliance, which noisily worked outside the legal-regulatory system to raise rank-and-file consciousness about Seabrook, embarrass political and economic elites, and challenge the wisdom of nuclear power. The Clams were not just against unchecked development; they opposed centralized political and economic power in general, which they believed Seabrook represented.

Long Island's opposition had a less "global" outlook than SAPL, far removed from the theories and methods of the Clamshell Alliance. But there were occasional demonstrations (many quite large) at Shoreham. Unlike Seabrook, however, no Long Islanders ever tied themselves to a Shoreham crane to get arrested (see Bedford 1990, 80–83). Although this Clamshell strategy did not necessarily win anti-Seabrook converts, it did put Seabrook opponents in a national light. Whether accurate or not, Seabrook was seen as *the* antinuclear movement of the 1970s in the United States. Shoreham's opposition rarely made headlines outside of Long Island and even had trouble getting noticed there.

L I L C O — *From Public Nuisance to Public Power*

On September 27, 1985, Hurricane Gloria ripped through the heart of Long Island, packing 110-mile-per-hour winds and leaving behind billions of dollars in damage. Besides damaging L I L C O's transmission grid, Gloria caused irreparable harm to the company's reputation. It was this unplanned act of nature that had the greatest impact on grassroots opposition to Shoreham and, increasingly, to L I L C O itself. A month after the hurricane, public opinion on Long Island had swelled against both the plant and the company. This was the first significant shift of public opinion in several years, far surpassing the impact of Three Mile Island or Suffolk's opposition to emergency evacuation. After Gloria, only 20 percent of Long Islanders favored opening Shoreham, whereas 71 percent opposed it. L I L C O chair William Catacasinos was "distrusted" by 65 percent of those polled.

This increased opposition was based on a growing mistrust of L I L C O's competency, and this growing mistrust was mostly due to Hurricane Gloria. As mentioned in the last chapter, those who had lost power were much more likely to bear negative feelings toward the company (*Newsday*, 17 November 1985, 3). Many Long Islanders were also incensed that L I L C O chair William Catacasinos did not rush back from his Italian vacation after the storm. Although this would have done substantively little to help L I L C O or its customers,

Long Islanders felt Catacasinos should at least bear some of the same inconveniences. Long Islanders started blaming every local problem on LILCO's managerial incompetence. Many Long Islanders started wondering whether LILCO had the organizational competence to operate a complex nuclear reactor.

In the eyes of many Long Islanders, LILCO effused organizational arrogance. Amid frequent requests for rate hikes (many of which were demanded by LILCO's creditors), LILCO continued to subsidize private pro-Shoreham groups, spent hundreds of thousands of dollars on a public relations campaign, and devised a generous set of golden parachutes for its top executives. William Catacasinos would receive $5 million if the company were taken over or if he retired. Another $5 million would be split among twenty-three other top executives (*Newsday*, 21 March 1987, 3). Catacasinos had something of a legacy in this regard. His former company, Applied Data Systems (Long Island), charged him with arranging parachutes that would pay him about $291,000 per year after he turned fifty-five (*Wall Street Journal*, 16 May 1984). This payment would have supplemented his LILCO retirement benefits. LILCO argued that this was the only way that top executives could be retained in a company constantly under stress (*Newsday*, 4 March 1987, 7).

LILCO's rank-and-file workers did not fare as well. After a month-long strike in July 1984—the company's first strike in its seventy-five-year existence—workers faced an eighteen-month pay freeze on top of a 5 percent pay cut and massive firings (one thousand jobs) unilaterally imposed earlier that year. Up to that point, LILCO employees had been the last bedrock of unwavering support for the company and for Shoreham (*Newsday*, 31 March 1984, 3; *Wall Street Journal*, 11 July 1984, 14 August 1984). Utility jobs were always viewed as lifetime employment with above average compensation, and workers had always defended the company's integrity. LILCO faced eroding employee support over the next several years.

An eclectic group of Shoreham and LILCO opponents began calling for a public takeover of the utility. One poll showed almost 70 percent of all Long Islanders favored a public takeover of LILCO (*Newsday*, 28 July 1986, 4). Ironically, this movement may have inadvertently sown the seeds of its own destruction. During debate over the one-dollar deal to close the plant, there was a lot of rhetoric about a public takeover, but little actual action. In fact, the way public power played out may have prevented independent, grassroots opponents from participating meaningfully in the discussions. The Long Island Power Authority (LIPA), created by the New York State government to explore a public takeover of LILCO, may have done more to eliminate public power as an alternative than any other player. This irony was rooted in the relationship between Shoreham's unofficial and official opposition and will be explored fully in the following section on "Nuclear Power and Official Opposition."

Seabrook—Treading Water[3]

The Three Mile Island and Chernobyl accidents did not propagate massive waves of anti-Seabrook sentiment. As with Shoreham, Three Mile Island did make relevant the emergency evacuation of Seabrook, which became the key grassroots issue by the mid-eighties. The eventual potency of this issue moved the anti-Seabrook discussion beyond abstract economic philosophy and complex nuclear technology, to whether you could escape the beach and protect your children during a nuclear accident (Bedford 1990, 87). Unofficial intervenors continued to participate in the evacuation hearings and were continually blamed for delays at Seabrook.

SAPL and a more subdued Clamshell Alliance used this issue to build anti-Seabrook sentiment, not an easy task with a limited budget and powerful opponents. The Clams did at times display the same creativity and ingenuity that contributed to their almost-legendary status in the national antinuclear community. For example, they understood some of the interorganizational relationships discussed in Chapter 4 and actually picketed some of the Boston banks that financed Seabrook. Grassroots opponents also tried to build anti-Seabrook fervor by painting PSNH as a corporate monster. PSNH was never a popular company. The utility had spent a great deal of money on Seabrook, continually demanded rate increases, and tried to overturn New Hampshire's CWIP law. When the law was first discussed in the late seventies, PSNH claimed it would ruin them financially while simultaneously implementing an enormous dividend increase. Many people were not amused (see Bedford 1990, 106).

PSNH was able to cloak its role in Seabrook with the formation of New Hampshire Yankee (NHY) in 1984. Now, Seabrook would be run by the independent NHY, which took directions from twelve owners throughout New England. All top NHY executives were brought in from other companies, and one of their strategies was to distance themselves from PSNH as much as possible. Former NHY president Ed Brown joked that the company's letterhead tried to keep its subsidiary relationship to PSNH in very tiny print.[4] This may have been a shrewd managerial move to distract grassroots opposition to Seabrook, but this alternative was possible only because of the unique circumstances of this plant. LILCO may or may not have pursued such a strategy. The point is moot because it was the sole owner of Shoreham and could not construct a parallel to New Hampshire Yankee. Shoreham and LILCO were a package deal.

New Hampshire Yankee executives believed that this and other strategies were somewhat successful in containing public opposition to Seabrook in the mid-eighties. According to their polls, opinion was about equally divided between supporters and opponents during this period, with proximity to the plant making a big difference. The company held focus groups with residents and learned that technical issues ("the plant is safe") did not work and that energy depletion warnings ("brownouts are coming") were only mildly successful in selling the reactor. Focus groups responded better to warm, personal messages from plant employees and their families, who said they lived close to the plant and had every reason to make sure it worked correctly.[5] LILCO could not use this particular strategy either, not because they did not think of it, but because of poor labor relations following the 1984 strike. Were New Hampshire Yankee and PSNH demonstrating shrewd management or were they merely exercising an obvious option which happened to be available to them but not to LILCO? And why were anti-Seabrook groups so ineffective in countering this campaign whereas anti-Shoreham forces successfully publicized the arrogance and ineptitude of LILCO? Were anti-Shoreham forces simply smarter and better?

Shoreham's grassroots opponents did not seem to have any organizational or financial superiority over Seabrook's opponents in the prelicensing period, and public opinion for and against each plant was fairly similar. In fact, in many ways, SAPL and Clamshell folks seemed to have a better grasp of complex issues, whereas Shoreham opponents were more rooted in a NIMBY philosophy concerned primarily with personal property values. For the sake of argument, let's say that the grassroots movements at each plant were equally effective or ineffective. Why, then, didn't "people power" shut down Seabrook? Clearly, "people power" alone does not explain the troubles in these two cases, or why Shoreham became the first licensed nuclear reactor to be shut down without ever operating.

Nuclear Power and Official Opposition

Suffolk County's and New York State's opposition to Shoreham was unprecedented in the history of U.S. nuclear power. Massachusetts joined this small club a few years later with its opposition to Seabrook. Such official government opposition is far more important than unofficial grassroots opposition for understanding why these similarly troubled plants had such different outcomes. There is a qualitative difference between grassroots opposition and government opposition. Structurally, governments have superior resources, including capital, labor power, and access to information. If Suffolk County or Massachusetts wanted a poll taken to evaluate people's response to an actual emergency, they could do it; if they needed to counter a LILCO or NHY ad campaign, they could do it. The coalitions of anti-Shoreham and anti-Seabrook activists had far fewer resources. Ideologically, government involvement can lend legitimacy to a movement that challenges powerful organizations and conventional wisdom. It may be harder to dismiss elected officials' objections to a nuclear reactor than to dismiss objections from environmentalists or other unofficial groups.

But not all official opposition is the same. Shoreham's official opposition was stronger, more timely, and more legitimate than Seabrook's. Shoreham's official opponents—Suffolk County and New York State—were formidable political and economic forces. Suffolk alone has a slightly larger population than the entire state of New Hampshire, and New York State spends more money in one year than New Hampshire does in thirty-two. Both Suffolk and New York State entered the fray in 1983, at the beginning of emergency evacuation hearings, although the state became a major player only later. The official opposition also added to the relatively strong legitimacy already enjoyed by the unofficial opposition. The county government was generally respected, and Mario Cuomo was New York's most popular governor since Nelson Rockefeller.

Seabrook's official opposition was far more anemic. There was simply no local equivalent to Suffolk County, and New Hampshire stood squarely behind the plant. When Massachusetts finally became involved in 1986, the evacuation hearings were well under way. Massachusetts did add significant resources to the federal regulatory battle over emergency evacuation, but had no jurisdiction in state and local battles over rate hikes and taxes. Massachusetts's intervention was also viewed by many (although not all) New Hampshirites as political opportunism by Michael Dukakis, or as meddling by bored Yankee liberals. Seabrook may not have opened if its official opposition were as powerful as Shoreham's.

Official Opposition at Shoreham

At the AEC construction permit hearings in 1971, then–Suffolk County executive H. Lee Dennison said that Shoreham's power would be crucial to the continued growth and development of Long Island (see McCaffrey 1991, 100). But, Suffolk County's political leaders grew more skeptical about the merits of atomic power on Long Island as electricity demand leveled out and Shoreham's costs increased astronomically. In August 1977, the county became an official intervenor in the NRC construction hearings on LILCO's proposed Jamesport nuclear plants.

The legislature grew more concerned with Shoreham in the wake of Three Mile Island and changing NRC regulations. In 1982, it fell one vote short of officially opposing the completion and licensing of Shoreham. The legislature was uncertain whether Long Island would need this extra power and if emergency evacuation was even possible. In abstaining from this vote, legislator Louis Petrizzo warned his colleagues that they were seriously jeopardizing the continued economic growth and development of Long Island (*Newsday*, 27 June 1979, 19).

In the next few years, the county and LILCO continually sparred about the ability to evacuate Long Island in case of a Shoreham accident. LILCO's Matthew Cordaro argued that the NRC's new ten-mile evacuation zone was too large and that a two-mile zone was more realistic (*Newsday*, 2 March 1982, 26). Suffolk was not convinced and continued examining demographic and topographic data. In July 1982, a county survey showed that 25 to 50 percent of *all* residents of *both* Suffolk and Nassau counties (approximately three million people) would evacuate in a Shoreham emergency (*Newsday*, 16 July 1982, 31). By December, another study showed that 75 percent of the bus drivers who were supposed to help evacuate schools would tend to their own families rather than carry out this duty. Suffolk still believed that evacuation was possible but that a twenty-mile evacuation zone was more acceptable (*Newsday*, 3 December 1982, 7). LILCO disagreed.

On February 17, 1983, the Suffolk County legislature voted 15–1 to withdraw its already-prepared evacuation plan for the Shoreham plant (*Newsday*, 18 February 1983, 3). This $600,000 plan, designed in June 1982, called for evacuating 744,000 people in a twenty-mile zone around the plant—instead of the ten-mile zone recommended by the NRC (Zorpette and Fitzgerald 1987). County Executive Peter Cohalan said that the unique conditions at Shoreham made evacuation impossible and that LILCO had made a blunder building the plant in such an area (*Newsday*, 17 February 1983, 3). LILCO disagreed with the county, arguing that NRC rules gave Suffolk no choice in the matter—it had to develop a plan (*Newsday*, 9 March 1983, 3).

New York State remained virtually uninvolved in Shoreham throughout the seventies and early eighties. After Suffolk County refused to participate in evacuation plans, Governor Cuomo appointed a blue ribbon factfinding commission to examine the issue in more detail. On December 18, 1983, the Marburger Commission released its report. John Marburger, president of the State University of New York at Stony Brook and a physicist, chaired this diverse group of New York businesspeople, medical experts, and grassroots activists. The 290-page document had ten points of unanimous agreement covering 3 pages, and 287 pages of disagreement. Marburger said that few new facts came out in the study, although lots of opinions did (*Newsday*, 19 December 1983, 3).

In addition to agreeing that the plant was a "mistake" and that LILCO lacked credibility, the commission advanced a startling hypothesis: that abandoning the plant, although expensive, was not necessarily equivalent to suicide for Long Island (*Newsday*, 19 December 1983, 3). LILCO had always argued that closing the plant would be catastrophic. The commission's conclusion drastically changed the parameters of discussion and gave the Cuomo administration some reason to reevaluate its position on Shoreham. Before the report, Cuomo had said he would not impose an evacuation plan on Suffolk (substituting state workers for county workers), but would work with the county to develop a plan if they wanted (*Newsday*, 18 February 1983, 4). Cuomo also suggested that the county not totally rule out the plant if it could somehow be opened safely. Economically, the governor had believed that closing the plant would be much more expensive for Long Island citizens than opening it (*Newsday*, 19 February 1983, 10). After the Marburger Commission's report, Cuomo began more forcefully opposing Shoreham's licensing (*Newsday*, 14 January 1984, 10).

By April 1984, Cuomo opposed any emergency drill, claiming that it was impossible to evacuate Long Island, and pressured the NRC and FEMA not to waste their time (*Newsday*, 25 April 1984, 5). He also assigned one of his top aides, Fabian Palomino, to represent New York in all NRC deliberations on Shoreham. Cuomo's opposition to the plant increased when a study by the State Energy Department's Office of Planning concluded that the electricity produced by Shoreham would not be needed until 1997 (*Newsday*, 6 December 1984, 7). LILCO disputed this and predicted brownouts during the upcoming summer without Shoreham on line (*Newsday*, 13 December 1984, 5).

The Cohalan Shuffle

During his May 1985 shift in position (see Chapters 3 and 4), Suffolk County executive Peter Cohalan said that new scientific evidence suggested that smaller evacuation zones were needed, and that no one would gain by continued opposition (*Newsday*, 31 May 1985, 5). The county legislature was not amused and promised to lead the anti-Shoreham opposition. Even Lou Howard, the sole Shoreham supporter in the legislature, thought Cohalan was out of line (*Newsday*, 1 June 1985, 4). The Cuomo administration continued to oppose a LILCO evacuation plan and assured Suffolk that it would never impose a plan on the county. A Long Island business leader said this may have been Cohalan's attempt to shift all evacuation responsibilities onto the state—a shift Cuomo wanted no part of.[6]

Many Shoreham opponents felt betrayed by Cohalan, because he had more or less assumed command of the "official" anti-Shoreham forces. Cohalan had been reelected in 1983 primarily on his anti-Shoreham position. Democratic state assembly member Patrick Halpin opposed Cohalan and campaigned that the county should stop wasting money opposing Shoreham, because its opening was inevitable (*Newsday*, 22 September 1983, 3). Cohalan's surprisingly narrow victory was probably due to his staunch position on Shoreham (*Newsday*, 9 November 1983, 5). Cries of "Peter the Fox" (Cohalan's middle name was Fox) became common, as 57 percent of surveyed Long Islanders opposed the Cohalan shuffle. Not surprisingly, this was about the same percentage (56 percent) that opposed Shoreham's opening (*Newsday*, 2 June 1985, 7).

This episode marked a critical change in Shoreham's official opponents. Until 1985, Suffolk County was clearly leading the fight against Shoreham, with the state an important, vocal supporter. But toward the end of 1985, Governor Cuomo began strengthening his opposition to Shoreham and to LILCO. Long Island Association president James LaRocca, a former state energy and transportation official under Cuomo, believed this was part of an evolutionary shift by the governor. In LaRocca's view, Cuomo eventually became "absolutely committed to not opening Shoreham," contributing significantly to the eventual outcome.[7] New York State's usurpation of the county's leadership became indisputable as the official opposition turned its attention to a public takeover of LILCO.

Public Power—Reality or Rhetoric?

In July 1986, Suffolk County and New York State created formal mechanisms for a takeover of LILCO. The Suffolk legislature voted 17–1 to float $7.5 million in bonds to raise money for a takeover (*Newsday*, 29 July 1986, 3). The state legislature established the Long Island Power Authority (LIPA) to examine this scenario. LIPA was officially created in January 1987 to initiate a public takeover only if rate savings were evident and only after negotiations with LILCO proved unsuccessful (*Wall Street Journal*, 28 July 1986). LILCO challenged LIPA as unconstitutional, but this position was eventually dismissed by a state court (LILCO 1987).

In September 1986, the State Court of Appeals ruled that only the state (and not Suffolk) had jurisdiction over a public takeover, because LILCO served more than two counties—Nassau, Suffolk, and Queens (*Newsday*, 6 September 1986, 1B). In one last gasp to stay involved, Suffolk sued LILCO under the Racketeer-influenced and Corrupt Organizations (RICO) Act for continually lying to the NYPSC about Shoreham's completion in order to obtain rate increases (*Newsday*, 4 March 1987, 7). With treble damages possible under a RICO lawsuit, LILCO could be liable for about $8.7 billion.

Cuomo announced in 1984 that he was a great believer in public power and had no problem with LILCO "taking a bath" over Shoreham (*Newsday*, 15 February 1984, 3). The governor also condemned any talk of bailing LILCO out of its past mistakes (*Newsday*, 24 October 1985, 3) and opposed any public takeover that included paying for Shoreham (*Newsday*, 22 May 1986, 4). In January 1986, Cuomo appointed a commission to investigate a public takeover of LILCO (*Wall Street Journal*, 31 January 1986). This group reported in June that a takeover would probably result in electric rates 7 to 9 percent lower than those charged by LILCO (*Newsday*, 23 June 1986, 3). Given these figures, Cuomo urged the legislature to create LIPA.

The law called for Cuomo to appoint five LIPA members, with Assembly Speaker Mel Miller and Senate Leader Warren Anderson appointing two each (*New York Times*, 11 January 1987, 1[XXI]). Miller selected longtime Shoreham opponent Irving Like and Stephen Liss, a top aide of assembly member Paul Harenberg who introduced the LIPA bill. Anderson, a Shoreham supporter, chose Martin Bernstein and Leon Campo. Bernstein was a Long Island real estate attorney and Hempstead (Long Island) town council member. Campo, head of the East Meadow (Long Island) school board and an increasingly visible Shoreham/LILCO critic, was a surprise selection. Campo's views on public power were unclear but he seemed to lean in that direction. Cuomo chose Long Island banker John Kanas, developer William Mack (to be LIPA chair), Shoreham Opponents Coalition's Nora Bredes, State Economic Development Director Vincent Tese, and State Consumer Protection Board President Richard Kessel.

The State Assumes Command

Cuomo's true commitment to public power was always questionable. The president of the Long Island Association, a close acquaintance of the governor, said, "Public power was sort of a leap-frog. People started talking and then others started talking. The governor never really wanted this. I mean who would want to own LILCO? Then, all of a sudden, there's a lot of activity, and there's LIPA. Then [the Cuomo administration] had to spend a whole lot of time to get it not to happen."[8] Camps were almost immediately established within LIPA about the feasibility of a public takeover of LILCO. Nora Bredes, Stephen Liss, and Irving Like all supported a takeover, with Leon Campo leaning that way. Martin Bernstein and John Kanas were opposed to a takeover. Bernstein said that "if you need a haircut, you don't have to buy the barbershop" (*Newsday*, 22 October 1987, 25). Kessel, Mack, and Tese—all Cuomo appointees—were initially unclear about their positions, but Mack and Tese quickly began advocating a negotiated settlement rather than a public takeover. Kessel was clearly torn between his origins in the grassroots anti–Shoreham/LILCO movement (which favored public power) and his allegiance to the governor, who was trying to get public power "not to happen."

It took LIPA four months to commission its first study and about nine months to retain full-time consultants. In September 1987, Lazard Frères issued a preliminary "verbal" report that LIPA rates following a takeover would be 7 to 10 percent cheaper than LILCO rates.

LILCO disputed this, claiming that LIPA rates would actually be 12 to 24 percent higher than LILCO rates (*Newsday*, 22 September 1987, 7). However, Lazard Frères never issued a written report, and LIPA came under increasing pressure from LILCO (to supply better evidence) and from grassroots supporters of public power (to get moving). At a December 1987 board meeting, the Nassau-Suffolk Neighborhood Network presented LIPA with a large pink plastic snail crawling toward a seven-foot-tall, smiling lightbulb (*Newsday*, 12 December 1987, 7).

The Cuomo administration's commitment to public power was also challenged when Tese, Mack, and Kessel commenced separate negotiations with LILCO while LIPA was supposedly deliberating a public takeover. Many critics felt this was an obvious conflict of interest that made a mockery of the governor's supposed commitment to public power. However, in his following State of the State address Cuomo reiterated his support of LIPA and public power (*Newsday*, 5 January 1988, 3). Leon Campo—temporarily aligned with Bredes, Liss, and Like—accused Mack and LIPA of stalling takeover proceedings to facilitate the negotiated settlement. Mack responded that LIPA would not move toward a takeover "until it was ready" and denied any stalling. Kessel, showing signs of accepting the governor's parameters on public power, patronized his critical board members and called the takeover "a very complex issue" (*Newsday*, 23 January 1988, 7). Kessel also said that LIPA should take a "prudent" approach to a takeover despite the Lazard Frères report of cheaper rates with public power (*Newsday*, 6 August 1987, 27). The following week, Mack repeated earlier statements that a settlement was preferable to a takeover (*Newsday*, 30 January 1988, 11), and a February LIPA meeting was actually canceled because it conflicted with these talks (*Newsday*, 26 February 1988, 21). Paul Harenberg, LIPA's legislative sponsor, was appalled and said he did not intend LIPA to become a watchdog to ensure that LILCO trimmed its trees (*Newsday*, 30 January 1988, 11; 22 March 1988, 2). Governor Cuomo believed that these public power advocates were misinterpreting the LIPA mandate, and

he continued using his LIPA appointees as envoys in "secret" negotiations with LILCO (*Newsday*, 29 April 1988, 3).

From Anti-LILCO to Anti-Shoreham

Cuomo's feigned commitment to public power was also reflected by
a shifting emphasis away from "anything but LILCO" to "anything
but Shoreham." Tese, Mack, and Kessel repeatedly said that closing
Shoreham was the key to any future actions, although LIPA's mandate was not that specific. With Suffolk County legally prohibited
from pursuing public power, and with grassroots organizations long
since institutionalized into the official process, there was little effective opposition to this redefinition except for Liss, Bredes, Campo,
and Like. Interestingly, just as the state/LILCO negotiations bogged
down and public criticism of LIPA swelled, Lazard Frères announced another "preliminary report" on the merits of a public
takeover. This draft concluded that a friendly takeover of LILCO by
LIPA would save $2.5 billion in rates over the next fifteen years. A hos-
tile takeover would cost more but would still significantly decrease
electric rates (*Newsday*, 28 March 1988, 6). A final draft of this report
was never seen, although public power advocates had been antici-
pating it for over a year.

Given these new figures, on March 29, 1988, LIPA finally made a
friendly offer to buy LILCO at $8.75 per share, about 10 percent be-
low its market value (*Wall Street Journal*, 30 March 1988). Kessel announced that this was a credible offer, and Tese canceled a negotiating session with LILCO to show his seriousness (*Newsday*, 30
March 1988, 3). However, Al Mazzorano of Dean Witter said that this
situation resembled professional wrestling because LIPA was under
pressure to do something, but knew that this offer would certainly be
tossed aside (*Newsday*, 30 March 1988, 3). In mid-April, the LILCO
board predictably rejected this offer (*Newsday*, 16 April 1988, 3). The
following week, LIPA raised its offer to $10 per share (*Newsday*, 26
April 1988, 5). In early May, the bid was raised to $12. Financial ana-

lysts considered these serious offers that LILCO had to consider (*Newsday*, 26 April 1988, 5; 6 May 1988, 3).

The One-Dollar Deal—Shoreham's Opposition Fragments

The Cuomo administration's abandonment of public power or of a LILCO takeover was cemented when the one-dollar deal was announced a few weeks later. Cuomo seemed to think there would be rejoicing by all of Shoreham's opponents—the wicked plant was dead! But the governor underestimated the depth of anti-LILCO sentiment on Long Island and within the Suffolk County government. Long Island rate payers, both residential and commercial, were not convinced that closing the plant was impossible without 63 percent rate hikes and a financially robust LILCO.

LILCO's customers had absolutely no say in this deal and had been represented only marginally in LIPA. With grassroots organizations long subsumed by official opponents, there were few noninstitutionalized routes for expressing displeasure. Thus the Suffolk County government and the New York State legislature spearheaded opposition to this deal, although in different ways. The county government played up its RICO suit as testimony to possibly defeating LILCO but had absolutely no veto power over the one-dollar deal. The state legislature, however, did have to formally approve the deal. LIPA also had to approve the deal officially, but its vote was never in question.

Supporters of the deal insisted that it would be best for Long Island's long-term economic growth and development. When this rhetoric failed to sway many opponents (unofficial or official), the Cuomo administration formed "truth squads," which would straighten out the misinformed and ignorance-based fear about this excellent deal (*Newsday*, 5 July 1988, 5). The Cuomo administration had already begun personal attacks on the deal's opponents. For example, when a utility analyst retained by Suffolk County argued that a LIPA takeover was cheaper than the settlement, Cuomo said that he must have used astrology to get his figures (*Newsday*, 10 June 1988, 5). Earlier, Tese had said that the analyst offered "nothing but smoke" (*Newsday*, 13 May 1988, 3).

The rhetoric increased as opposition to the deal intensified. In early July, Cuomo led the truth squads to Long Island on a whirlwind tour of local media and public gatherings. He argued that opposing the deal was synonymous with opening the plant. Opponents, he argued, were extremists teeming with misinformation, while the "responsible middle ground" supported the deal (*Newsday*, 5 July 1988, 5). The governor presented the deal as "take it or leave it" (*Newsday*, 7 June 1988, 3), and Vincent Tese said he was not at all interested in renegotiating (*Newsday*, 27 July 1988, 7). Richard Kessel asserted that all opposition was "irresponsible" until a better alternative was arranged (*Newsday*, 17 July 1988, 3), and Cuomo said that the state legislators from Long Island were "scared" to approve the deal because of the rate hikes (*Newsday*, 15 July 1988, 7). After this deal temporarily fell through in December, one of Cuomo's top aides actually said that the opposition "called people names and engaged in acts while we did not" (*Newsday*, 2 December 1988, 3).

Despite Cuomo's enormous popularity, political clout, and ability to constrain certain agencies, neither Long Island businesses nor its residents were convinced that ten 5 percent rate increases were in their best interests (*Newsday*, 17 July 1988, 3). When Kessel claimed that "there's a debt on Shoreham and someone has to pay for it" (*Newsday*, 19 June 1988, 4), not everyone thought that Long Island rate payers had to be that someone. Soon, the Cuomo administration could or would frame any Shoreham-related decision only in terms of this agreement and how it was the best solution to a difficult problem. In this fascinating twist, a once-stalwart official opponent to Shoreham and LILCO had become one of LILCO's saviors.

Official Opposition to Seabrook

The brevity of this section shows the difference in official opposition at Shoreham and Seabrook. Although Seabrook and its owners faced a relatively formidable and effective grassroots opposition, it did not have to worry much about more powerful official opposition. There was official government opposition to Seabrook from a number of sources, but it paled compared to Shoreham's official opposition. If Seabrook's owners had faced more official opposition, the reactor might not have opened; and without the resistance of Suffolk County and New York State, Shoreham might have opened.

Many of the townships near Seabrook "officially" opposed the plant at one time or another, but they did not have very significant resources against the largest utility in the state. Seabrook is located in Rockingham County which, like all New Hampshire counties, has very little economic or political influence. Only the state of Massachusetts offered any significant official opposition to the plant. Like New York, Massachusetts refused to impose an emergency evacuation plan on its towns that fell within the NRC evacuation zone. Massachusetts was not involved other than this significant regulatory obstruction. According to NHY president Ted Feigenbaum, the weakness of the surrounding towns and counties, coupled with home state support, was crucial to opening the plant.[9]

The only possible source of official opposition within New Hampshire was the state itself, and it turned out to be one of the most ardent supporters of Seabrook, especially during John Sununu's two terms as governor. An engineer by training, Sununu was a vociferous advocate of nuclear power and feared that closing Seabrook might destroy the nuclear industry. In the seventies, Governor Meldrim Thompson was equally supportive of Seabrook and railed against anyone who did not support Seabrook's construction (Bedford 1990, 72). Judd Gregg, who followed Sununu as governor, successfully derailed talk about a possible public takeover of PSNH during bankruptcy hearings (Bankruptcy Document #3404).

There was, however, one minor breakdown of this historical support of Seabrook. Thompson was defeated for reelection in 1978 by Democrat Hugh Gallen. Thompson's defeat rested almost strictly on his veto of the legislature's Construction While in Progress bill

(CWIP), which Gallen quickly signed. As I discussed earlier, this was an extraordinarily important event for PSNH and greatly affected the complex relationship among PSNH, its creditors, and the state Public Utilities Commission. This bill did not by itself reflect any anti-Seabrook sentiment, and Gallen assured PSNH that he was not opposed to the plant (Bedford 1990, 106–107). At certain moments in the eighties political elites became angry with PSNH, but this never approached systematic opposition. Most recently, PSNH annoyed Governor Gregg during bankruptcy deliberations when it announced its desire to be regulated in the future by the Federal Energy Regulatory Commission (FERC). This would have eliminated New Hampshire's role in rate approval (through the NHPUC) and allowed Seabrook's costs to be passed on to rate payers sooner, because FERC did not have a CWIP rule. New Hampshire Yankee executives felt this was a misguided decision by PSNH, but it did not seem to generate any substantial official opposition.

Assessing Official Opposition

The wishes and desires of Long Islanders did not by themselves close Shoreham. Seabrook's grassroots opponents had the same wishes and desires, and that plant is on line. In addition, neither Long Islanders nor New Hampshirites wanted the mandatory rate increases resulting from the one-dollar deal and the Northeast Utilities takeover of PSNH. Why, then, was there so much official opposition to Shoreham and so little to Seabrook? One possible explanation is that representative democracy is more alive on Long Island than in New Hampshire. Perhaps the symbiotic relationship between elites and masses has not matured in the Granite State.

But this one-dimensional explanation ignores some of the more subtle social elements that contributed to the different levels of official opposition. First, Long Islanders' public opinion on Shoreham generally followed rather than preceded elite decisions. Second, Suffolk County's government was unusually centralized even by New

York standards. Neighboring Nassau County had a far less centralized system, where five popularly elected town supervisors selected
someone as Nassau county supervisor. These six people constituted
the Nassau County Board of Supervisors, which convened but had
very limited power or resources. Instead, individual townships were
the locus of power in Nassau County.

Suffolk once had a township style of government but switched to
the more centralized form in the 1970s.[10] This shift had nothing to
do with Shoreham. Had Suffolk maintained its former system there
would have been far less salient official opposition to Shoreham. The
towns of Brookhaven and Riverhead, which were affected by the
evacuation plans, might have offered some resistance but not nearly
as much as Suffolk. Even so, these two townships have more economic resources than their analogous townships in New Hampshire.
New Hampshire's local political structure simply did not allow for
any significant official opposition below the state level. This had
nothing to do with "people power."

These distinct political configurations explain only part of the differences in official opposition. There was also something very different about Suffolk County that fomented this sort of official opposition,
above and beyond any personal political idiosyncracies. The unprecedented official opposition to Shoreham was caused by unique
political-economic circumstances. Local governments do not generally act this way.

Nuclear Power and Local Growth Coalitions

The different outcomes at Shoreham and Seabrook were greatly affected by the specific morphology of local growth paradigms and the
institutional cohesion around these paradigms. Long Island and
southeastern New Hampshire exhibited an interesting potpourri of
urban, rural, and suburban characteristics—although the latter two
were clearly dominant. In both cases, there was a lack of ideological

consensus about the proper paradigms for stimulating local economic growth and development, and about how a nuclear plant fit into these competing paradigms. This lack of an ideological growth consensus was mirrored by the absence of strong, unified growth coalitions which could powerfully influence local social policy — especially concerning the two nuclear plants.

But these two growth coalitions were like night and day when compared to each other. Relatively speaking, Long Island had a weak local growth coalition, whereas southern New Hampshire had a strong one. There are four ways this local political-economic context patterned the resolution of Shoreham (including the one-dollar deal). First, Long Island's extremely weak growth coalition provided fertile ground for official opposition to Shoreham. Second, this weak coalition exaggerated the apparent effectiveness of "people power" in opposing the plant. Third, LILCO's allegedly horrible mismanagement also seemed exaggerated because of this unique context — which left LILCO isolated in the local corporate community. Finally, this weak local coalition allowed an outside coalition — in this case, the Cuomo administration — to force the one-dollar deal on local elites and nonelites.

Conversely, a relevant and powerful official opposition to Seabrook was all but impossible given the local political-economic context in New Hampshire. Unofficial opposition was left on its own and faced a much more formidable "pro-Seabrook" contingency. PSNH was not isolated or abandoned by its corporate brethren, so its equally questionable management did not seem as horrible. This section examines the relative state of growth coalitions on Long Island and in New Hampshire, and how these situations affected Shoreham and Seabrook.

Competing Growth Paradigms on Long Island

On the surface, Long Island has a confusing political-economic landscape. It is both urban and rural, yet it is neither urban nor rural. It is quite suburban but also lacks many suburban qualities. In short, Long Island does not fit within standard sociological categories of place. Thirty years ago, Long Island was New York City's biggest "bedroom community," while also maintaining its agricultural heritage. More recently, it has developed a personality of its own, including independent manufacturing and service companies and some large financial firms. Even though the Long Island Expressway—the main artery to Manhattan—is called "the world's longest parking lot," only about one-quarter of Long Islanders work in New York City, and their number is continually decreasing (*Newsday*, 25 January 1988, 7). Suffolk County was actually New York's largest agricultural community (in terms of revenues) with $92.2 million in sales in 1982 (Zweig 1987).

The town of Southold has become an interesting metaphor for competing growth paradigms on Long Island. Southold is on eastern Long Island's North Fork, a heavily agricultural area with some summer tourism and a permanent population that traces its roots to the seventeenth century. More than seventy miles from Manhattan, Southold was never part of the Long Island bedroom community and developed independently from its "up island" natives. But, in the last decade, the urban/suburban sprawl of Long Island has reached Southold, and it has become a contested terrain because of different visions of local economic growth and development.

As suburban development—either residential or retail—encroached, falling potato prices forced some small farmers out of business, and skyrocketing land prices lured many others away. Upscale houses proliferated as farms were carved up into two acre lots (Zweig 1987). Potato fields and sod farms are dotted with occasional patches of townhouses and condominiums. Many farmhouses sport for-sale signs and rusty tractors parked by dilapidated barns.

> The North Fork is confronted with a choice. It can easily go the route of residential and commercial development so typical of the history of Western Long Island. Or it can preserve its agricultural and rural char-

acter. There are extremely powerful forces operating to turn the region into a residential area with associated commercial districts and a [large] tourist trade drawn by the surrounding waters. But there is nothing inevitable about this outcome. (Zweig 1987, 39–40)

This may not be an inevitable outcome of such growth battles, but it has become the most likely outcome throughout the United States. The prevailing definitions of land use that we explored in Chapter 1 favor "modern," residential/commercial definitions over "old-fashioned" agricultural (and conservationist) ones. Frequently, these are class struggles in which wealthy or powerful people succeed at the expense of the less powerful. Low-income housing is not among the visions of most Southold developers. However, Long Island—especially Suffolk County—has shown enormous resistance to these conventional definitions of growth and how these definitions are implemented. Although some of its rural base has been permanently lost, eastern Long Island has not become overrun with video stores and grotesque townhouses. Despite powerful forces, not everyone agrees that condominiums and Taco Bells are the only way to stimulate Long Island's economy.

Institutional Growth Coalitions

There are two components of local institutional growth coalitions: one being the level of cohesion within the local business community around any existing or emerging growth paradigms; the other being the strength and predictability of local business-government relations around these growth paradigms. In Chapter 1 we saw how strong growth coalitions (or "growth machines") limit the parameters of acceptable social policy. These policies often benefit certain classes and organizations whose members do not necessarily participate in policy making. There was no such institutionalized growth coalition on Long Island. Long Island's corporate community was extremely fragmented. In addition, the Suffolk County government was famous for challenging conventional paradigms of growth and development even before its unprecedented opposition to Shoreham. In addition to lacking a mature business community, Long Island lacked a *corporate* community. While more and more corporations were moving to Long Island throughout the sixties, seventies, and eighties, Long Island's capitalist base was oriented to small business, agriculture, and tourism. The structural coalitions that would have supported Shoreham's opening were essentially corporate in nature, but there was no corporate base on Long Island, and the ever-increasing corporations have still not formed a corporate community.

The Long Island Association[11]

Formed in 1926, the Long Island Association (LIA) has grown into the largest business group in Nassau and Suffolk counties. Historically, the LIA represented primarily small business interests, because Long Island was home to predominantly small firms. As the island's corporate community expanded, the LIA increasingly reflected the interests of bigger companies. However, the organization could never reconcile the differences between its small business base and its growing corporate presence. This internal conflict between big and small firms, as well as the lack of large local investment banks, contributed to the formation of a very decentralized business community. LIA president James LaRocca said that, in order to get Long Island's "power structure" into one place, you would need a building the size of the Nassau Coliseum (capacity 16,000). In contrast, he said, when you visit certain elite "lunch clubs" in Rochester or Buffalo, you can see 90 percent of the local power structure at the table.

After Suffolk County announced its opposition to Shoreham in 1983, LIA president Walter Osterbrink said that the organization had favored opening the plant in the past but now would begin discussing other alternatives (*Newsday*, 23 February 1983, 3). Still, three days later, Osterbrink was lobbying Governor Cuomo that, because abandoning the plant would be catastrophic, it should be opened quickly

(*Newsday*, 26 February 1983, 3). Executive Vice President William Gaylor agreed that abandoning the plant would catapult electric rates such that no one would be able to do business on Long Island (*Newsday*, 27 February 1983, 3). By the end of 1984, the LIA had no consistent position on Shoreham, and Osterbrink said he saw no immediate resolution of the situation (*Newsday*, 23 November 1984, 23).

The Grumman Corporation, an LIA member, started talks with Georgia officials about possibly moving some of its operations south. This came in the wake of LILCO's proposed 56 percent rate hike in 1983. Grumman was by far Long Island's largest company, employing 22,000 workers on a payroll of $600 million, most of which went directly into the local economy (*Wall Street Journal*, 15 July 1983). Grumman was spending about $25 million a year on electricity out of its total $2 billion costs. The company was impressed with Georgia's significantly lower electric rates.

Others in the business community, however, felt that Grumman was overstating its problem and was merely trying to save some money with veiled threats of abandoning Long Island. Chemical Bank's Jerry Gilbert said that electricity costs were only a small portion of total business costs on Long Island and that the region would already be an economic wasteland if these costs were so important (*Newsday*, 27 February 1983, 3). Whether or not electricity costs really were a problem, it is interesting to note the lack of agreement within the Long Island business community on this issue.

The disagreement intensified in January 1984, when LILCO offered a 5 percent rate discount for its three thousand largest commercial customers. Despite repeated warnings about brownouts, LILCO claimed that this would use "unsold electricity" (*Newsday*, 25 January 1984, 3). Later that year, LILCO made an additional deal with Grumman, its largest customer, which eventually saved the company about $2 million a year in electricity costs (*Newsday*, 3 June 1985, 1B). Small businesses were not eligible for these discounts and began calling for cost controls on LILCO prices (*Newsday*, 20 May 1984, 5; 7 January 1985, 5). Timothy Archdeacon, president of the Long Island

Restaurant and Caterers Association, called these cut rates nothing more than a pacifier (for Grumman) and was skeptical that members of this small business group would be eligible for discounts.

Long Island's small business community began voicing its opposition to Shoreham (because it would increase rates even more) and complaining about corporate domination of the LIA (*Newsday*, 3 June 1985, 1B). Ira Leibowitz of the Regional Energy Action Coalition (a small business group opposed to Shoreham) claimed that "the LIA is not really pushing very hard for the smaller companies [and have] written them off" (*Newsday*, 3 June 1985, 1B). Southern Container's Vice President Steven Hill felt that "the LIA represents Grumman and LILCO [not] firms like ours," and Arthur Metzger of Amco Plastics asserted that the LIA did not include small companies in the vital discussions of the issues (*Newsday*, 3 June 1985, 1B).

Walter Osterbrink defended the LIA, claiming that it would have better represented these issues if its members had spoken more loudly about them (*Newsday*, 3 June 1985, 1B). Director of Legal and Economic Affairs Jim Faraldi said, "You don't have to ask them all the time what they think. They are members. They join for what we stand for" (*Newsday*, 3 June 1985, 1B). Despite these avowals, the LIA was becoming more and more dominated by Long Island's growing corporate community. By the middle of 1985, only one board member—Santos Abrilz of Apoca Industries—represented smaller businesses. The other seats were filled with representatives of Newsday, Citibank, Hazeltine, and Roosevelt Raceway (*Newsday*, 20 September 1985, 25). Former State Transportation Commissioner and Energy Commissioner James LaRocca became LIA president and the organization's foremost leader. LaRocca's position on Shoreham was clear when, as energy commissioner, he made the plant part of his "master plan" for electricity in New York (*Newsday*, 18 July 1985, 81). However, LaRocca's insistence that LILCO's evacuation plan be tested did not resolve the internal LIA dissension about Shoreham and LILCO.[12]

The LIA was also plagued by a lack of individual and organizational leadership. Grumman's strength, coupled with the local commercial banking community's relative weakness, prevented the LIA from forging any coalitions around Shoreham or other connected issues. Grumman's rhetoric about electricity rates demonstrated a "businesswide" interest, but its actions favored only Grumman. For example, as Grumman struck its deal with LILCO to obtain cheap hydropower from upstate New York, it also began building its own 25-megawatt generator to further reduce its electricity costs and dependence on LILCO. Grumman president John Bierwirth said that opening Shoreham was not as important as "the economics and availability of power for all Long Island" (*Newsday*, 3 June 1985, 1B). Yet, Grumman eventually enjoyed $2 million in savings while LILCO *raised* many other commercial rates to balance the revenue losses from Grumman's growing energy independence (*Newsday*, 25 November 1985, 23). Grumman's discourse about Long Island's energy needs was not reflected in its generally parochial actions.

This is not to imply that Grumman should or could have orchestrated some classwide rationality among the Long Island political economy. Perhaps financial organizations are better able to coordinate policies that benefit the business *community* rather than individual businesses. Long Island did not have a mature investment banking community. Instead, local capital usually flowed through smaller savings and loans or satellite branches of New York City's commercial banks. For one brief moment in 1985, the local financial community was positioned to take the LIA's reins. When Walter Osterbrink retired as LIA chair, he was immediately replaced by Richard Dalrymple, executive vice president of European American — the island's largest commercial bank (*Three Village Herald*, 2 May 1985). A few months later, Dalrymple had to vacate this post when he was fired from European American following some financial problems at the oft-plagued bank (*Newsday*, 20 September 1985, 33). Perhaps if Dalrymple had stayed in this LIA leadership position, Long Island's fledgling financial community would have been able to mend some of the rifts within the Long Island business community, where big companies argued with small, where Grumman showed little concern for its corporate brethren, and where LILCO was quickly becoming a black sheep. Only *Newsday* publicly supported Shoreham's opening and supported LILCO. No one else seemed quite sure what to think.

Long Island's Missing Growth Coalition

The Association for a Better Long Island (ABLI)—a group of developers not directly tied to the LIA—issued a 1987 statement condemning the Suffolk County government's historic role in developing environmentally conscious legislation. ABLI member Jack Kulka said that Suffolk County was becoming an increasingly unfriendly place to do business. "The local government should be doing more about garbage and less about halting [housing and office] development which increases building costs," he said (*Newsday*, 26 January 1987, 3C). In fact, the county government would soon do something about Long Island's garbage problem (mandatory recycling), but the ABLI wouldn't like that either. It was just another in a long history of policies that challenged the allegedly universal benefits of unlimited, corporate-oriented economic growth.

In 1970 Suffolk became the first local municipality to ban phosphate-based detergents. This triggered legislation in many other areas so that, by 1981, most detergent companies no longer used these environmentally unsafe ingredients anywhere in the United States. In 1981 Suffolk enacted bottle and can deposit laws, a movement that soon spread throughout the country. Three years later, the county banned smoking in all public places, and its law was used as a model for other U.S. antismoking legislation. Opposition to Shoreham was not the county's first unprecedented move.

The Long Island business community did not embrace these laws, but thought of them as mere inconveniences rather than impediments to local economic growth and development. More troublesome was the county's program of buying up the undisturbed pine barrens throughout eastern Long Island. These protect Suffolk's water supply which is totally dependent on natural aquifers. Much of this land was eyed by developers who imagined condominiums and golf courses instead of pine trees. A growth paradigm concerned with increasing upscale housing units does not complement a strategy of preserving natural areas. Suffolk's desire to limit this kind of development was one of the most serious institutional obstacles to building a more conventional local growth coalition. Reflecting this contradiction of growth strategies, Long Island's business community was torn more over public land acquisition (and the taxes financing it) than over any other issue except Shoreham. The LIA as an organization supported land acquisition, which upset many of its members. The ABLI emerged partly as a response to this position.

Conventional politics in Suffolk defied most standard two-party generalizations. The legislature frequently displayed coalitions of Democrats and Conservatives, and the Republican presiding officer, Greg Blass, won support from Democrats and named one of them as deputy presiding officer. In this body, the location of constituents was much more important than party affiliation. Blass and fellow Republican Fred Thiele represented Long Island's eastern sections. Both legislators had been catalysts in the county's "antigrowth" orientation — reflecting eastern Long Islanders' concern with the environment. As presiding officer, Blass had additional organizational resources to pursue these policies. "Controls are needed to protect Suffolk's quality of life and its water supply, the very things that make the county a sought habitat," he said. "There have to be limits to growth. After all, we are an island economy . . . and without water, there is no island" (*Newsday*, 26 January 1987, 3C). LIA president James LaRocca was also not thrilled at the county's unpredictable nature. "Suffolk County is repeatedly identified as an area where they are likely to legislate in any area, no matter whether it is needed or not," he said. "It creates a perception that the county is likely to create unique local laws and burdens on doing business that exist nowhere else" (*New York Times*, 1 July 1990, B1).

Business Opposition to the One-Dollar Deal

Leaders of the Long Island business community were not convinced that it was in Suffolk County's best interests to assist LILCO's return to financial health (*Suffolk Life*, 25 May 1988). The LIA initially supported the deal (*Newsday*, 13 May 1988, 3) and earlier had indicated that a negotiated settlement would favorably end all the confusion over Shoreham; it was also far superior to the "socialist" concept of public power (*Newsday*, 8 February 1988, 7). *Newsday* publisher Robert Johnson tried forging some classwide rationality by offering to act as a sounding board among all parties to the deal, because it was certainly "good for economic growth" on Long Island (*Newsday*, 17 May 1988, 4). The paper's editorial position was also that the agreement was excellent for growth and development (*Newsday*, 15 May 1988, 5).

But Newsday would soon be the lone local corporate supporter of the deal. A month after the deal was cut, the LIA announced that it was opposed to the agreement and about 80 to 90 percent of its members were opposed. LIA chair John V. N. Klein announced that the deal was better for LILCO than for the rest of Long Island's business community (*Newsday*, 25 June 1988, 11). The LIA cited the rate increases as harmful to Long Island's continued economic growth, because businesses would not be able to afford operating there (*Newsday*, 3 July 1988, 30). The ABLI opposed the agreement on similar grounds (*Newsday*, 24 June 1988, 7). Individual corporations also took exception to the deal. Grumman claimed that the electric rate increases would add $250,000 to the cost of each plane it manufactured, leading to layoffs and a significant drop in local economic growth (*Newsday*, 25 November 1988, 5).

Ironically, just when the Long Island corporate community showed signs of coming together on this issue, its influence was usurped by a different coalition. In fact, the increasingly cohesive Long Island Association was not particularly concerned with LILCO's credit rating (*Newsday*, 24 June 1988, 7). It seemed that, besides Newsday, the only Long Island corporation left that cared about LILCO was LILCO. But, under the direction of the Cuomo administration, local businesses were not involved in forging or approving the deal. A state-level growth coalition had moved into Long Island's "growth vacuum."

The New York State Growth Coalition

The financial community had always displayed a certain skepticism about Cuomo's commitment to public power. Salomon Brothers' Mark Luftig said that Cuomo left himself obvious outs by including the "lower rates" and "must negotiate" clauses (*Wall Street Journal*, 8 January 1986). Daniel Scotto of L. F. Rothschild commented that institutional investors were not likely to take LIPA seriously, and Shearson-Lehman's Rodd Anderson speculated that Cuomo had no desire to go into the utility business (*Wall Street Journal*, 16 March 1988). These analysts seemed to be correct.

Vincent Tese, wearing two hats for Governor Cuomo, made an initial proposal to LILCO in early 1987, before LIPA had even seriously considered a takeover strategy. This deal offered LILCO financial solvency in return for closing down Shoreham (*New York Times*, 24 May 1987, 1[XXI]). LILCO turned it down and made a counteroffer to the state which, among other things, offered to sell Shoreham for one dollar (*Newsday*, 18 October 1987, 5). Tese did not immediately respond, but called for new talks between the state, LILCO, and the investment banks representing each of them. "If the bankers get a clear indication that LILCO is serious about negotiations, we'll tell them to proceed" (*Newsday*, 22 October 1987, 25). The introduction of these financial participants was more than coincidental, and bank representatives often met without Tese and Catacasinos present (*Newsday*, 7 November 1987, 3). Political-economic consensus about LILCO and Shoreham was now being orchestrated at the state level. The financial community clearly believed that an *immediate* solution was far better than any distant solution. Although there was no consensus on whether a takeover or a negotiated settlement would be substantively more advantageous, either would be acceptable as long as it happened quickly and ended the uncertainty around Shoreham, LILCO, and Long Island's electricity. William Sans of First Pacific said that he wouldn't mind seeing a takeover at around $15 per share. "I think obviously the ratepayers or [the state] should pay for Shoreham and not the shareholders. It's not their fault . . . you know" (*Newsday*, 22 March 1988, 2). Long Island's commercial and residential rate payers did not agree with this assessment. But did that really matter?

Growth Paradigms and Coalitions in New Hampshire

There was by no means an absolute consensus about "correct" economic growth and development in southern New Hampshire. Mostly rural and small-town, New Hampshire could not avoid Boston's suburban sprawl of the seventies. Southern New Hampshire became both a Boston bedroom community and a corporate haven, with both groups attracted by the state's practically nonexistent taxes. As on Long Island, there was ideological conflict between these newer paradigms of growth and development, which were interested in building condominiums and malls, and more established paradigms centered on agriculture, natural resources, and tourism. Seabrook was couched in the former paradigm. New Hampshire's ever-expanding population and economy would require more electricity. According to some longtime New Hampshire residents, the Seabrook plant symbolized the fast-paced growth and commercialization they opposed (*Boston Globe*, 8 November 1988).

A 1972 poll by PSNH tapped into this ideological conflict. Half of all New Hampshire residents approved of nuclear power as an environmentally friendly energy source, and about half thought that building Seabrook was a good idea. However, when asked whether they preferred economic growth or environmental protection, more than 70 percent chose the former, and only 11 percent felt that New Hampshire's economic growth was too slow. This preference was true especially among younger people and seacoast residents (Bedford 1990, 66–67). Because all of Rockingham County has only about 250,000 people, this sort of ideological dissension was not quite as potent as similar dissension on Long Island. More importantly, there was little doubt where New Hampshire's business community (and its growing corporate community) stood on economic growth and development and on opening the Seabrook plant.

The Business and Industry Association (BIA), was New Hampshire's equivalent to the LIA.[13] But, the BIA was far less internally divided than the LIA. This seemed mostly due to the domination of the BIA by New Hampshire's largest corporation and its largest commercial banks. Even though corporate capitalism was slow arriving in both New Hampshire and Long Island, it arrived more forcefully in New Hampshire. The state's relative corporate maturity allowed the BIA to operate in a more unified, cohesive ideological framework. Thus, in New Hampshire, the business community spoke with a clearer, more consistent voice on Seabrook. It also helped that, unlike LILCO, New Hampshire Yankee remained active in the BIA and on generally good terms with other members. New Hampshire Yankee executives said that BIA support was critical in successfully opening the plant.[14]

The strength of this corporate community was reflected in, among other things, the vociferous support of the *Manchester Union-Leader*, the state's largest and most influential newspaper. In addition to clear, unwavering editorial support for Seabrook and PSNH, the paper's news stories provided glowing accounts about Seabrook while subtly ridiculing the reactor's opponents (Bedford 1990, 120). On Long Island, *Newsday* generally supported LILCO and Shoreham editorially, but its news coverage was, by mainstream standards, thorough and unbiased, if not particularly creative in seeking out unofficial sources. The Boston media, on the other hand, were far less supportive of Seabrook. For instance, some Boston television stations actually offered equal time to Seabrook opponents after a barrage of New Hampshire Yankee ads in 1987 (*Boston Globe*, 12 March 1987).

This may also have added to some New Hampshirites' insistence that the anti-Seabrook forces were just a bunch of Massachusetts liberals.

The corporate-based cohesion of New Hampshire's business community, and the utter lack of such cohesion on Long Island, was clearly evident during a series of PSNH rate hike requests in 1986 and 1987. In 1986 the utility asked for a 10 percent hike, whereupon Anheuser-Busch, The Nashua Corporation, and Digital (the state's second largest employer) all warned that such an increase would make them leave New Hampshire. Anheuser-Busch also said that it was looking into producing its own electricity (*Boston Globe*, 20 July 1986). The request was almost immediately scaled back. The following year, amid a series of prebankruptcy rate hike requests, PSNH announced a five year electricity discount for The Nashua Corporation, which had threatened to relocate in West Virginia. The BIA's Ian Wilson applauded the move, saying, "This is what we have been saying for years: keep rates low and companies will stay" (*Boston Globe*, 2 December 1987). Within the next few weeks, twenty-four other large New Hampshire corporations had approached PSNH for similar rate discounts (*Boston Globe*, 24 December 1987).

Businesspeople and outside analysts acknowledged that this would increase the rate burden on small companies and nonbusiness customers. This did not seem to concern the corporate-dominated BIA. Henry Lee, energy specialist at the JFK School of Government, said that the idea of large companies' producing their own power is frightening to utilities (*Boston Globe*, 2 December 1987). Small businesses and nonbusinesses do not usually have this particular option to leverage rate reductions, a point seemingly unimportant to the BIA. It is interesting to compare these events to the Grumman-LILCO electricity deal on Long Island. There, the organizational selfishness of one huge company exacerbated the organizational tensions within the LIA and within the local business community. In New Hampshire, the corporate-oriented BIA thought that the PSNH-Nashua agreement was terrific, and it seemed to provide a framework in which other large corporations could extract similar discounts. On the whole, it was not just certain New Hampshire corporations that benefited but also the New Hampshire corporate community. The corporate capitalist growth paradigm clearly dominated organizational dynamics in the New Hampshire business community.[15]

The organizational strength and influence of the BIA was also apparent during PSNH's bankruptcy hearings. The BIA opposed rate hikes under PSNH's reorganization plan, claiming that it would cost 22,600 jobs throughout New Hampshire over five years (*Boston Globe*, 8 March 1989). Although BIA's data collection was highly problematic, the report still influenced PSNH. When Northeast Utilities entered the reorganization fray, BIA eventually supported its reorganization plan, even though it called for slightly higher rates than PSNH's most recent plan (*Boston Globe*, 14 November 1989). BIA's logic, according to New Hampshire Yankee executives, was that slightly higher rates were a small price to pay for the stability, predictability, and certainty that Northeast would provide. The BIA thought a Northeast takeover would be better for the state's long-term economic growth and development.[16] This was a sterling example of classwide rationality by BIA, in which the parochial interests of one longtime member were sacrificed for the overall benefit of the entire corporate community—at least as BIA defined these interests. Interestingly, all of this influence occurred after BIA was officially removed from bankruptcy hearings by Judge Yacos (Bankruptcy Document #2677).

BIA helped New Hampshire Yankee maintain a positive image of Seabrook in the eyes of other New Hampshire corporations. There was some opposition to Seabrook within the BIA, but this opposition never affected BIA's *organizational* position on either Seabrook or its convictions about New Hampshire's economic growth and development. More importantly, BIA was "absolutely instrumental" in maintaining healthy relations between New Hampshire Yankee and the state of New Hampshire through the long delays at Seabrook.[17] This might not have made much difference under the friendly watch of John Sununu, but it seemed to help the state legislature to remain committed to Seabrook despite opposition from some representatives near the plant. A less organized, less effective, less influential BIA may not have been as helpful. These political-economic circumstances reduced or eliminated the likelihood that any effective official opposition would emerge against Seabrook, regardless of local "people power."

Institutional Coalitions in New Hampshire

Shoreham certainly would have opened if John Sununu had been governor of New York. Sununu would have been happy to develop a state-run evacuation plan for Shoreham, as he did for Seabrook, and the legislature probably would not have stopped him. However, Sununu's presence in New Hampshire was not by itself sufficient to open Seabrook, although it certainly helped. It was Sununu who originally conceived of New Hampshire Yankee (although not the name) in order to facilitate completing and opening Seabrook (Bedford 1990, 119). Sununu was seen as so important to Seabrook's future that, during the 1984 debt restructuring, Robert Hildreth said that the refinancing might fall through if Sununu were not reelected that fall (*Boston Globe*, 15 September 1984). Sununu had no problem removing from the NHPUC an outspoken Seabrook critic and replacing that person with a vociferous Seabrook supporter (*Boston Globe*, 6 July 1987).

Nevertheless, Sununu came very close to losing a 1986 reelection bid to Paul McEachern, an attorney representing Hampton Town before the NRC. McEachern ran on a strict anti-Seabrook platform, promising to rescind state support for the emergency evacuation plan if elected. Even as a Democrat in an overwhelmingly Republican state, McEachern just narrowly missed victory (*Manchester Union-Leader*, 1 February 1988; Bedford 1990, 152). This would appear to be a strong popular outcry against the plant, but it seemed to have no effect on the state government as a whole. By contrast, in Suffolk County it was impossible to win any county election after 1984 without a firm anti-Shoreham position. At one point, there were at least ten different local political parties, each in some way opposed to Shoreham, LILCO, or both.

Sununu, meanwhile, continued as one of Seabrook's biggest boosters, even as he was heading to Washington, D.C., as George Bush's chief of staff. But even without Sununu's eager support, the New Hampshire growth coalition was firmly behind Seabrook. State police and courts were notoriously strict in punishing civil disobedience against the plant and often used what Clamshell leader Guy Chichester called "violent rhetoric" concerning the use of dogs, horses, and other painful instruments against protestors (Bedford 1990, 78). Small businesspeople as well as nonbusinesspeople were amazed at how ineffective the legislature was and how corporate interests seemed to dictate both the dialogue and the actions of state elites (Bedford 1990, 86, 89). It is unclear whether this growth coalition operated in the second or third dimension of power. A second-dimensional view would see key organizational and class interests wielding enormous power (usually covert) to achieve and maintain a political economic climate that serves their own vested interests. A third-dimensional view would see these biased outcomes embedded in the entire logic of the local political economy, with certain class and organizational interests being served even without participation (overt or covert) by these agents. In all likelihood, both perspectives are accurate. Most importantly, they go far beyond "people power" and other first-dimensional explanations for the very different outcomes at Shoreham and Seabrook.

Conclusion

The romanticized vision of "people power" gets heavy play in explanations of Shoreham's closing, as Judge Weinstein's opening quote clearly illustrates. The extreme version of this position often takes on David and Goliath tones—where the wishes and desires of everyday citizens combine with mom-and-pop businesses and dedicated local public officials to slay the fire-breathing corporate behemoth despite its industry and federal government benefactors. This romantic vision is not only incorrect but dangerous as well. It is incorrect because Shoreham's closing (assumedly a victory) was only one part of the $1 deal which was filled with losses for the Davids and made Goliath stronger than it had been in years. This "victory" already leaves Long Islanders with the highest electricity rates in the United States with another seven years of mandatory rate hikes yet to come. Romanticizing "people power" is dangerous because it obfuscates the powerful structural constraints in which everyday people must live their lives, and diverts attention from the individual, organizational, and class interests which are frequently served at their expense.

If people power was the key to Shoreham's closing, then Seabrook also should have closed. Seabrook's grassroots opposition was certainly equivalent in size and scope to Shoreham's and in many ways could even be considered more effective. But there was no effective official opposition to the Seabrook plant that provided critical structural and ideological resources, in turn exaggerating the effectiveness of grassroots opposition. I do not mean to belittle the very important activities of unofficial opponents, but overestimating "people power" runs the risk of underestimating the significant obstacles preventing people from making history more to their liking.

The emergence of official opposition in Suffolk County was not rooted in the personality characteristics of its local elites or in the vibrancy of representative democracy on Long Island but in local political-economic peculiarities. The absence of any local growth consensus or growth coalitions provided a fertile medium for this unprecedented official opposition to a nuclear reactor and a large local utility. This growth vacuum also reduced the potential effectiveness of Shoreham's defenders, who had few of the political and economic resources usually associated with nuclear plant supporters. Unfortunately for Shoreham's and LILCO's opponents, this vacuum, coupled with the institutionalization of all unofficial opposition, allowed a more cohesive state growth coalition to take control and guarantee a mostly victorious outcome for LILCO's creditors and large institutional shareholders.

New Hampshire, in contrast, had a relatively strong local growth consensus and growth coalition which did not help spawn official opposition. It also provided more significant and powerful adversaries for Seabrook's grassroots opponents. This local growth coalition was guided—consciously or not—by the principle that what is good for centralized, corporate capitalism is good for everybody. Without such a powerful local growth coalition, it is possible that Seabrook would have gone down the tubes along with other abandoned nuclear plants in recent history. As Mr. Feigenbaum's opening quote illustrates, Seabrook had the "right" support.

CHAPTER SIX

NUCLEAR POWER, SOCIAL POWER,

AND DEMOCRACY

■

There are a lot of ordinary people in this thing, and we're not making the decisions, and they're crucifying us.
—Albert Mall, with $70,000 invested in LILCO stock (*New York Times*, 24 August 1984)

You have to trust government officials since the average person doesn't have expertise. Progress always involves risk.
—Randy Scheller, Stony Brook, New York (*Newsday*, 9 October 1983)

When I make a mistake, when I mess up somebody's hair, nobody else pays for it, I pay for it.
—Long Island hairdresser (*Newsday*, 2 March 1989)

At the beginning of this book, I outlined my reasons for studying Shoreham and Seabrook, reasons that are both academic and personal. Academically, I am offering what may be a unique analytical framework for studying the many faces of social power. By combining ideas from different areas of sociology, puzzling phenomena may become more understandable. This eclectic analytical framework should prove useful for studying more than nuclear plants. Personally, my research has been driven by my judgment that these cases violate the spirit if not the letter of a democratic society. This judgment has nothing to do with nuclear power per se; the cases could have been about convention centers, office buildings, or football stadiums. The Shoreham and Seabrook stories just happen to exemplify how important decisions are made unfairly and at the expense of individuals and groups who have very little power over the decisions.

Thus my two reasons for pursuing this research complement each other. I believe that this integrative theoretical approach will provide a mechanism for recognizing and analyzing the deleterious effects of unequal social power, especially when this power is obfuscated by democratic rhetoric. In the first section of this brief concluding chapter, I will underscore some of the important theoretical ideas employed in this tale of two nukes. In the second section I will suggest that this theoretical perspective can and should identify what Mills (1959) called threats to reason and freedom. The "promise" of sociology, Mills insisted, was providing insights that would help us participate in decisions affecting our lives.

Theories as Tools

Some sociologist once said that theories were like tools and that we would be most successful explaining human behavior if we used the right tools for the right job. This seems sensible but both graduate education and the profession itself often contradict this simple intimation. There is often enormous emphasis on theoretical specialization, best illustrated by questions like, "Are you an organizations person?" Methodological compartmentalization is also encouraged (often called the "quantoids" and the "squishes"). Those of us who prefer a more diverse approach may be ostracized from all camps. The "organizations people" will call us dogmatic Marxists for being interested in social class; the social class camp will call us "Weberian" for looking at corporations and organizations. We may never get another party invitation.

But, there are many interesting social phenomena that cannot be explained by rigidly adhering to one theory or one approach. If we really want to explain complex events and not just demonstrate the prowess of certain theories, we should use all the necessary tools. Imagine building a house with just a hammer or stir-frying some General Tso's chicken with a single chopstick. Shoreham was an interesting phenomenon in 1983. The Suffolk County government had just withdrawn from the emergency evacuation plan and had begun opposing the plant's construction. This was unprecedented behavior for a local government and seemed to be a surprising affirmation of the pluralist definition of procedural democracy in which public preferences are reflected in elite policies. But as the story unfolded, it became evident that this pluralist notion of democracy was descriptively useful and analytically useless. Despite the relatively strong unofficial and official opposition on Long Island (and eventually from New York State), construction of the plant moved ahead. Government policy makers opposed construction, but it seemed as if their opposition was irrelevant. The action was not taking place in the easily observable first-dimensional arenas of electoral politics and interest groups.

More complex ideas in political sociology went beyond these first-dimensional limitations, but still offered unsatisfying explanations for Shoreham. Many of these ideas did not work well when applied to the local, non-urban politics which were most salient at Shoreham. There was also the more general insistence that all relevant social power is located in political arenas. Ideas in economic sociology filled in many of these gaps but also brought similar problems. These ideas examined how corporations and organizations execute class interests (see Perrucci and Potter 1989, 6–7), but often treat political institutions as a vague, monolithic entity.

Synthesis: Local Political Economy

Many of these ideas from political sociology and economic sociology seem to complement each other and provide a theoretical approach that can unravel the Shoreham quandary. This approach considers political activity as important as economic activity, local events as important as national events, and nonurban areas as important as cities. The nonurban, local political economies of Long Island and southern New Hampshire were very different and led to the different outcomes at Shoreham and Seabrook. This suburban twist on the basic growth machine idea offers a more fulfilling analysis of Shoreham and Seabrook. Theories of corporate and intercorporate structure make it possible to avoid treating the "corporate community" as some monolithic black box. Finance hegemony theory provides a great tool for organizationally dissecting LILCO and PSNH, and explaining why local growth coalitions do or do not form. But even here a little twist is necessary. The capital controlled by banks is important, but other sources of external capital—such as bonds, stocks, and institutional investment—also demand attention. The organizational components of Shoreham and Seabrook begin making even more sense with this expanded definition of external capital.

These diverse tools work together to provide a more satisfactory explanation of these two nuclear plants and the nuclear industry. I believe that this theory of local political economies can also be employed to examine other similar quasipublic projects which are purported to serve general social interest while really serving more narrow interests. For instance, suppose we wanted to explore how and why a spanking new stadium (with dozens of luxury boxes) was built in a Florida suburb in order to lure a professional sports team from Cleveland to relocate in the Sunshine State. We might start by examining how the new stadium was financed (internally, externally, publicly, privately) and what sorts of public subsidies were offered by the local government to the future franchise owners. Was this subsidy the result of a public referendum? If so, what sort of growth and development rhetoric was used to justify the expense? Did the subsidy arise because of direct pressure from the franchise owners or other organizations which stood to gain? Was the subsidy a third-dimensional response of the local government which, without pressure from the benefiting individuals, internalized the logic of corporate capitalism to transfer public money to private hands? There is no reason to assume that the answers to these questions are exclusive of one another. Indeed, it is quite likely that social power is being exercised in all dimensions and includes individual, organizational, institutional, and ideological elements. The more flexible we are in our choice of tools, the more satisfying our analysis.

Theoretical Middle Grounds

I also hope this local political-economic framework mitigates certain polemical arguments within sociological schools. For instance, the so-called instrumentalist versus structuralist debate within political sociology has been the source of many heated exchanges and has become a standard paradigm conflict in political sociology texts and classes. The argument revolves around how the state systematically serves dominant class interests.[1] The instrumentalist position claims that members of the dominant class capture key command posts within the state, which leads to biased decisions favoring their own class interests. The structuralist position insists that the state serves dominant class interests even without dominant class members' participating in government decision making. It may even be more advantageous for capitalists to stay out of these positions so the state can act as the autonomous executive committee of the ruling class.

The discussion in Chapter 3 plays right into this debate. The empirical evidence suggests that Shoreham and Seabrook are best explained by the structuralist perspective, but this is far from absolute. There were many instances in which the actual people involved, and the interests they either held personally or represented, did make a difference in the biased decisions of the government regulatory body. I have emphasized this structural regulatory bias because it is more congruent with the third dimension of power. But it seems more fruitful to use both of these tools rather than insisting that only one of them must be correct. I believe that both the instrumentalist and the structuralist position offer insights into Shoreham and Seabrook. For instance, the New York Public Service Commission and the

New Hampshire Public Utilities Commission certainly had a built-in historical bias favoring the regulated utilities. But actual decisions about Shoreham's one-dollar deal and Seabrook's bankruptcy reorganization may have hinged on the commissions' membership. Governors Cuomo and Sununu certainly believed that their appointments would help achieve their goals.

This nondichotomous theoretical approach also plays well when we examine the Atomic Energy Commission and Nuclear Regulatory Commission both in these specific cases and in the general history of civilian nuclear power. Peter Freitag's study of federal regulatory agencies challenges the instrumentalist view that regulatory bias emanates from agencies stocked with former, present, and future members of the businesses being regulated, that the "foxes have taken over the hen house" (1983, 489). This analysis is convincing given the agencies sampled. But the AEC/NRC offers a much cloudier picture. In the early days of the civilian program, the agency was loaded with members of the industry—a clear source of bias. However, as the industry matured, this membership bias declined significantly. In fact, the rage since the mid-eighties has been to fill the NRC with former nuclear submarine commanders. It is not immediately evident that these people represent the class interests of those who benefit from a strong civilian nuclear program.

The water grows more murky if we consider commissioners like James Asseltine and Victor Gillinsky, who attest to the possibility that individuals on these agencies may not represent dominant class interests, or that individual actions might actually confront the structural bias of these agencies. If, however unlikely, there were one or two more commissioners cut from this cloth, NRC decisions about Shoreham and Seabrook may have been radically different. Similarly, I like to think that, if Governor Cuomo had asked a bunch of sociology graduate students to sit on the PSC in 1989, we might have made decisions that did not kowtow to the interests of capitalists, corporations, or corporate capitalism. Ed Royce (1993) warns social historians not to commit an ontological fallacy by treating "what was" as "what had to be." What might have happened is as theoretically interesting as what did happen.

Using elements of both instrumentalism and structuralism complements my central idea about the many faces of social power. From a second-dimensional view of power, identifiable actors can overtly or covertly manipulate regulatory agendas in order to serve personal, organizational (especially corporate), or class interests that contradict the publicly declared mission of the agency. From a third-dimensional view of power, membership in a regulated industry or the dominant class is not necessary for subscribing to the world view that government regulation should primarily be a boost to capitalists or corporate capitalism, or that the public interest is best served by subsidizing corporate capitalism and letting the benefits trickle down. I find this third dimensional view much more interesting because it explores the realm of ideology and "false consciousness"—when people not benefiting from growth coalitions become independent technicians serving growth coalitions.

Power as Constraint

Most sociology texts talk about power as "influence," "coercion," and "authority." Influence is persuasive power, coercion is forced power, and authority is legitimate power. All suggest that one entity's behavior can be affected by another entity's power—in whatever form it takes. This study of Shoreham and Seabrook offers a slightly different way to conceptualize social power. I like to think of "influence" and "control" as endpoints on a power continuum with "constraint" somewhere in the middle. The degree of short-term compliance by the less powerful social actor accounts for these categories' relative positions on the spectrum. When powerful actors exercise control, less powerful actors are highly compliant. When powerful actors exercise influence, less powerful actors are relatively less compliant. Constraint generates a medium level of short-term compliance.

I view authority as an important component of each type of power rather than some separate form of power. Because control often relies on coercion, legitimacy (authority) is not necessary for short-term compliance, although it may in fact be present. Indeed, power is often coercive in the absence of legitimacy. The irony is that coercive control is a tenuous strategy for *long-term* compliance. Coercion often breeds rebellion and defiance. For example, abusive parents often maintain a high degree of short-term control over their children, but they risk losing all power in the long run when the children leave home or, perhaps, kill them. The degree to which the children have legitimized their parent's cruelty will certainly affect this long-term relationship.

Influence, on the other hand, *requires* legitimacy in order to be successful. Compliance results when less powerful actors believe that more powerful actors have the expertise or the right to exercise power. For instance, teachers generally exercise power not through the coercive use of grades, but through the supposed expertise they have been socially granted. Grades become a primary mechanism of power only in the absence (for whatever reason) of a teacher's legitimacy. Outside the classroom, a local banker may have great influence over the local government's zoning ordinances because that person has been defined as an expert on economic growth. Once influence is successfully established, it is a relatively stable form of long-term power despite being unpredictable in the short run.

But social actors are not puppets which mindlessly obey powerful forces. Individuals always have some discretion. My idea of constraint emerges from the tension between compliance and discretion. Constraint features both a high degree of discretion by the weaker party and a high level of compliance with the wishes and desires of the powerful. This is also advantageous (to those exercising power) in the long run, because acquiescence is accompanied by the existence of alternatives for the less powerful, despite the limitation of these alternatives by the more powerful.

Thus powerful individuals, organizations, institutions, and ideolo-

gies constrain less powerful actors when they create, shape, and eliminate available options and alternatives. There may, in fact, be many
choices available in any given situation. But available choices (or
those perceived to exist) and unavailable choices have been leveraged by more powerful forces. Constraint is exercised by powerful
forces when they establish certain parameters and boundaries within
which less powerful forces make decisions. As a result, short-term
compliance is likely because only those options preferred by the powerful are immediately available. Long-term compliance is also prob-
able because of the exaggerated sense of discretion.

For example, some teachers only give students take-home exami-
nations with a selection of essay questions. These students probably
believe that they have a high degree of discretion over their answers.
But these exams can be constructed so that many of the questions
deal with the same substantive issue and differ only in how they are
phrased or presented. The teacher is assured of compliance with the
topics they want discussed, but the students believe they have
tremendous choice. If the teacher only provided one alternative
(control), the students might find the instructor and the course more
unreasonable, do lower-quality work, and make a fuss on evaluations.
If the instructor merely asked (influence) that the students address
certain topics, they might end up writing essays off the topic. So, in
order to ensure compliance and retain some sense of student discretion, cleverly constructed essay choices constrain alternatives while
appearing to provide great discretion.

The one-dollar deal at Shoreham and PSNH's bankruptcy illustrate constraint in action. As negotiations between New York and
LILCO concluded, two alternatives were publicly offered: either the
state (hence, taxpayers and rate payers) bailed out LILCO in ex-
change for closing the plant, or the plant opened. Period. Of course,
many other options were possible but they had either been removed
from consideration or never considered in the first place. For example, because LILCO is a public utility, the state could have confis-
cated its generating capacity, established a municipal power com-

pany, and left the owners and creditors gnawing on each other for their share of LILCO's assets. There may be extraordinarily good reasons why this alternative should not have been chosen, but these reasons were not explored. This option was never even considered, much less discussed. In New Hampshire, bankruptcy court (and the bankruptcy laws that inform it) could have constructed the receiving line differently so that rate payers and small stockholders got first dibs in PSNH's reorganization plan. Again, this may be a terrible idea, but such a case was never made explicitly. It did not really have to be made explicitly, because most social actors accept the conventional distribution of social power which ignores these options. Similarly, students might prefer not taking any tests or devising their own essay questions, but they rarely raise this option. It falls outside the socially constructed parameters in which they think and act.

Social Power and Democracy

This discussion of theory and theories is not simply an academic exercise. These theoretical ideas may provide an opportunity for regular people to live higher-quality lives because they can see how democratic rhetoric often distorts antidemocratic actions. Just because someone has choices, either on the election ballot or at the supermarket, does not mean that he or she is aware of or has access to ultimately preferable alternatives. Alternatives are restricted not necessarily by evil people allied in a conspiracy, but by subtle forces that shape and reproduce the unequal distribution of social power. What about the candidates and issues that never make it onto a ballot? Why was there no public referendum on the one-dollar deal or PSNH's reorganization plan?

Even the way we think about electricity production illustrates the subtle social constraint of alternatives. For example, truly public power is rarely discussed as a systematic alternative to centralized, corporatized, capitalist systems for producing and distributing electricity. This is caused neither by articulated public preferences (the first dimension of power) nor by the electric utility industry quashing this idea (the second dimension of power). Instead, it reflects an almost invisible a priori elimination of options before they are even considered (the third dimension of power). Public power is almost automatically seen as antithetical to the conventional definition of progress and proper economic growth: it is decentralized, noncorporate, and possibly socialistic. Innovative attempts at alternative energy production systems are frequently constrained by these socially constructed parameters. I recall a 60 *Minutes* segment some years back in which a company built thousands of solar panels in the California desert and was trying to transport this energy back to Los Angeles to compete with more conventional electricity producers. This company did not have much success because of the high cost of transmission. But I would argue that the shortcomings of this project were more conceptual than economic. For certain alternative technologies to be most effective, they must be decentralized and take simultaneous advantage of a variety of renewable sources (sun, wind, water, waste) instead of depending on just one. But such a decentralized conception of electricity production and distribution falls beyond the normal boundaries of thought, discussion, and action.

I am not advocating one form of technology or another. Nor am I condemning nuclear power as an evil that will destroy humankind. My intent is to debunk the idea that societies, and the people composing them, make decisions "naturally." Instead, decisions are made by people, and they are made from alternatives that have been constrained by the dynamics of social power. This power is exercised in ideological, institutional, organizational, and individual domains which frequently create and eliminate the awareness of or existence of policy options. This constraint of alternatives leads to biased decisions which benefit certain people, groups and classes at the expense of others. Most importantly, those who suffer most from these decisions have internalized the consequences as "normal" and, like Mr.

Mall in the opening quote of this chapter, often believe they have no alternatives except crucifixion.

Social Power and the Law

Conventional rhetoric often insists that democratic societies operate under the "rule of law" rather than the "rule of people." Universalistic, written laws can overcome exceptionalistic interests and ensure that all people be treated fairly regardless of class, physical characteristics, or beliefs. The judicial system that articulates and interprets these laws is frequently viewed as the guardian of democracy. Many of us feel that the chronic corruption of political and economic institutions has somehow eluded the courts. Certainly, many of the regular individuals and groups fighting unofficially and officially against Seabrook, PSNH, LILCO, and Shoreham believed that the federal court system would see the reasonableness of their case. Unfortunately, in these two cases the courts proved to be just another mechanism for reinforcing the unequal distribution of social power and ensuring that those individuals and organizations used to winning keep winning.

At Seabrook, as we saw in Chapters 3 and 4, less powerful people and organizations were categorically rebuffed in bankruptcy court. Bankruptcy laws were simply not designed to represent the interests of rate payers or small shareholders. Some people's optimism about PSNH's entering Chapter 11 was quickly reversed when the bias of bankruptcy law became clear. Judge Yacos's personality may have had some bearing on the case, but it seemed that the structure of Chapter 11 bankruptcy was far more important than the judge. Seabrook opponents thought it quite reasonable to wonder how a bankrupt company could operate a nuclear plant and evacuate parts of two states. But this issue simply was not relevant in bankruptcy court.

Shoreham, as usual, provides an even more interesting case of judicial bias. As I mentioned in Chapter 3, Suffolk County sued

LILCO under the federal Racketeer-influenced and Corrupt Organizations (RICO) Act for lying to the NYPSC in order to garner rate increases. During these hearings, it was hard to discern the line between Judge Jack Weinstein's personal biases and the structural biases of the law.

The trial began in late September, 1988. On December 5, only a few days after the first one-dollar deal fell through, the jury found LILCO guilty of fraud and awarded Suffolk $22.9 million in damages. LILCO planned to appeal but, due to an earlier ruling by Weinstein, also faced a class action suit on the same charges which could result in a multibillion-dollar verdict. Meanwhile, Judge Weinstein felt it within his judicial purview to order Suffolk and LILCO to attempt an out-of-court settlement. In January 1989, he appointed Kenneth Feinberg to mediate these discussions. Feinberg's initial proposal was that LILCO pay back rate payers approximately $400 million over the next ten years. Suffolk's objection was that Feinberg's thinking was being influenced by the components of the first one-dollar deal—which the county opposed and, more importantly, which had nothing to do with U.S. District Court or Judge Weinstein.

While Suffolk and LILCO haggled, Weinstein made it increasingly clear that he would like the matter expedited. He also indicated at this time that, although he believed the jury's verdict in the case, he would have reached a different conclusion. "I have never hidden my view that had I decided the case I would not have decided the same way the jury did. The county's use of the racketeering laws is one of the basic weaknesses of this case. It's one of the reasons this case does not have the value that some of the . . . attorneys believe it does" (*Newsday*, 3 February 1989, 3).

Then, on February 11, in an amazing twist, Weinstein dismissed the lawsuit ruling that RICO "does not apply to a rate regulation case such as the one before us," which belongs before the PSC rather than a federal court (*Newsday*, 12 February 1989, 5). Simultaneously, Weinstein strongly encouraged the parties to keep negotiating. "There is no better time than now for resolution of the entire controversy. Even this case could go on almost indefinitely through protracted appeals which can only benefit the lawyers *and put LILCO at further risk. Removal of the uncertainty is essential to the welfare of Long Island"* [emphasis added]. The judicial branch of government had now weighed in with its view of "proper" economic growth and development. There are no statutory restrictions on verdicts that might jeopardize a corporation's growth and stability, or statutory imperatives to link such growth to a local community's well-being.

Weinstein's decision did not invalidate the jury's verdict; it only said that the verdict (and the trial) should not have happened in U.S. district court. Any appeals, which Suffolk promised to file, would have to deal only with this jurisdictional question, not with the substantive verdict by the jury. With this move, LILCO could still face a guilty verdict immediately after an appellate court's ruling—another trial being unnecessary. The timing of this decision is also peculiar because from the beginning, LILCO's position was that RICO was not applicable. Weinstein *dismissed* this contention in May 1988 (*Newsday*, 18 May 1988, 5).

These blatantly contradictory decisions, and many subsequent actions, suggest strongly that Judge Weinstein very much favored some version of the one-dollar deal which had fallen through in December 1988. The class action suit of rate payers settled out of court a few days later, and was covered with Weinstein's fingerprints. Judith Vladek, appointed by Weinstein to represent the class in negotiations with LILCO, believed this settlement was in the best interests of all Long Islanders. Of the remaining eight class action representatives, seven disagreed. Fred Harrison, a schoolteacher, said he respected Vladek but was not enamored with the agreement. "[She] was under incredible pressure to settle from [Judge] Weinstein. I think she felt all along that the judge held a club to her head that if she didn't play ball, he wouldn't certify the class" (*Newsday*, 16 February 1989, 3). Judge Weinstein's maneuverings seemed like an attempt to make a settlement more likely by raising the potential damages all players faced if they continued holding out. As a result, it increased the likelihood of a second one-dollar deal almost identical to the first. It was never made clear exactly why Judge Weinstein supported the one-dollar deal or how his advocacy was in any way justified by the rule of law. It is also worth wondering if these decisions would have been different with another presiding judge, or if Weinstein merely reflected some embedded judicial logic which legitimates centralized, corporate capitalism. In any case, this is an excellent example of how the many faces of social power work to perpetuate dominant motifs of local economic growth and development which serve very narrow organizational and class interests.

Conclusion

While moving to Philadelphia in 1990, I checked the local yellow pages for the emergency evacuation plans at the nearby Limerick nuclear plant. I was pleased that the instructions included what to do with your pets because I was considering buying a Labrador retriever and did not want to misplace it during a core meltdown. These very sober evacuation plans ("turn off the lights") struck me as a good example of what C. Wright Mills called "crackpot realism." Mills was referring to certain countries "planning" how to survive a nuclear war rather than figuring out how to avoid one. Consider Dr. Strangelove, in the Stanley Kubrick movie of the same name, pondering over General Turgidson's suggestion that, if the United States launched a surprise nuclear attack on the Soviets, it would result in the loss of only 20 to 30 million Americans. "I'm not saying we wouldn't get our hair mussed up a bit," the general reassured those gathered in the war room.

The appearance of emergency evacuation plans in the yellow pages suggests that, like General Turgidson's war casualties, this is just the way things are. Or, as Mr. Scheller posits in this chapter's opening quote, that this is the normal risk associated with "progress." But nuclear power, like any other technology, is not just some normal progression of humankind; it is the result of a series of decisions made by a variety of people and organizations who may or may not have been concerned with interests beyond their own. Perhaps if we can see that "progress" is socially constructed rather than inevitable, we can also recognize that seemingly "natural" decisions are actually rooted in the dynamics of social power. Once we believe that important decisions affecting our lives are made by actual social actors rather than by magic, we are more likely to start demanding a more integral involvement with such decisions.

The key to a more democratic society, then, is first to illustrate the seemingly invisible social characteristics that lead many of us to think that our lives are inevitably determined and that we cannot do anything about it. For instance, there was little evidence at Shoreham and Seabrook to support Mr. Scheller's opening quote about trusting the government, even though most people on Long Island and in southern New Hampshire would probably agree with him. Perhaps a seed of skepticism can be planted in our minds that the so-called experts may not share our concerns. It is possible that such skepticism might be the first step in building a society that is democratic, not because people get to vote and write letters to their representatives, but because those who are affected by important decisions help make those decisions.

I would argue that our conventional views of social power and democracy are similar to our conventional views of electric power: they advocate centralization and corporatization. Somehow, we believe our society has improved because specialized political elites — the "experts" — now take care of important technical decisions while we sit back and enjoy life or struggle to survive. To establish a less centralized, less corporate formula of social decision making would indicate a regression in the normal course of human progress. In an Orwellian corollary, we might even say that democracy increases the less we participate in the decisions affecting our lives. Or, as one political scientist has said, the problem with democracy is that there's too damn much of it.

It seems that what goes around comes around. At random intervals since the mid-1980s, the nuclear industry has argued that new technologies have created inherently safe nuclear reactors that simply cannot melt down or leak (for example, see *New York Times*, 11 November 1987, D10). Mostly, the mainstream press parrots these claims (for example, see *New York Times*, 8 December 1989, B4). My father—who still lives on Long Island—has also been sending me newspaper clippings about LILCO. It seems that the new governor, George Pataki, is encouraging the vestigial Long Island Power Authority to sell $4.5 billion in tax-exempt bonds to purchase LILCO's entire electrical system (*New York Times*, 21 January 1996, B25). Much of this $4.5 billion will go to restructure LILCO's still-enormous Shoreham debt while supposedly lowering electricity rates by 12 percent. I wonder how the financial community is involved with this plan. I wonder why such a plan was less attractive six years ago, before the mandatory 63 percent rate hikes kicked in. I wonder why this allegedly probusiness governor is advocating a public takeover of a private company. I wonder, as did the hairdresser whose quote opened this chapter, who is going to be responsible if the powerful General Turgidsons of the world muss up our hair.

NOTES

Introduction

1. Jim Jasper (1990, xii) takes an almost-identical political position in his book *Nuclear Politics*. We developed this perspective independently, although I am glad to have Jim's company.

Chapter 1

1. See, for example, Edgar (1970); Hayes (1972); Ewen (1978); Ratcliff, Gallagher, and Ratcliff (1979); Ratcliff (1980); Whitt (1982); and Friedland (1983).

2. Shoreham and LILCO were regulated by the New York Public Service Commission (NYPSC). Seabrook and PSNH were regulated by the New Hampshire Public Utilities Commission (NHPUC).

3. Underwriting is little more than a promise of collateral for which banks collect large fees.

4. This gives a new meaning to "being Trumped."

Chapter 2

1. These include Gandara (1977), DelSesto (1979), Ford (1982), Hertsgaard (1983), Camilleri (1984), Mazuzan and Walker (1984), Clarke (1985), Campbell (1988), Hardert et al. (1989), Jasper (1990), and Walker (1992).

2. I heartily recommend watching the movie *The Atomic Cafe* for an up-close view of this nuclear progress package.

3. Sometimes multiple ASLBS, each dealing with a different substantive issue, are assigned to a case.

4. For more detailed information on the PRDP, see Mazuzan and Walker (1984).

5. These were the Yankee Rowe (Massachusetts), Columbus, NE, and Detroit Fermi plants.

6. This was the so-called Rasmussen Report. Details of this study can be found in most of the sources listed in note 1 (see especially Ford 1982).

7. Palladino worked in Westinghouse's reactor research labs for twenty years, was chair of Penn State's Department of Nuclear Engineering, and eventually sat on the congressional Advisory Committee on Reactor Safeguards (ACRS).

8. The Price-Anderson Act eventually sustained a judicial challenge and was ruled constitutional by the Supreme Court in 1977. The limit on liability was raised to $7 billion in 1988. Although this is more significant, it still falls far short of what an accident would probably cost.

9. A few states, like New York, allowed companies to recoup a small piece of these investments before a plant was completed. New Hampshire did not permit reimbursement for construction work in progress (CWIP).

10. Capitalization ratios are a simple measure of a company's assets and debts. This information is important to a company's potential investors and creditors.

11. U.S. Department of Energy, *Nuclear Power Plant Cancellations: Causes, Costs and Consequences*, DOE/EIA-0392 (Washington, D.C.: U.S. Government Printing Office, 1983).

12. These advantages, however, might be balanced by the less diversified risk.

13. Ironically, the 1973 oil embargo spawned a new perspective supporting atomic energy as a path to energy independence from countries in southwest Asia (see Gamson and Modigliani 1989, 15–17).

Chapter 3

1. Even though PSNH did not survive bankruptcy, the actual "owners" of Seabrook (especially its creditors) fared quite well. I will address this more fully in Chapter 4.

2. LILCO initially received its construction permit in April. After an appeal by intervenors, the permit became official in October.

3. I will discuss LHSG's role as "unofficial opponent" to Shoreham in Chapter 5.

4. Information in this section comes primarily from Bedford 1990, 35–46, and Stever 1980, 96–110.

5. Chapter 5 examines these two opposition groups in a little more detail.

6. Cuomo was wrong. The courts ruled this within the NRC's rights.

7. Letter from FEMA to NRC, 28 April 1987, NRC File on Legal and Adjudicatory Correspondence.

8. ASLB Hearings, 27 May 1988, 14 June 1988.

9. This relationship between internal capital (rates) and external capital (loans and so forth) will be discussed in Chapter 4.

10. NHPUC Ruling, 25 May 1978.

11. This draws from Ed Royce's (1993) idea of "constriction of possibilities."

Chapter 4

1. 1974 PSNH Bond Prospectus.

2. For instance, LILCO borrowed $5 million from Chase Manhattan in the early seventies—nothing compared to the billions LILCO would later borrow from a variety of lending consortia, of which Chase Manhattan was the lead bank.

3. New Hampshire Yankee interview, 5 June 1992.

4. The rate increase was approved only with some help from Governor Sununu (see Chapter 3).

5. See Royce (1993) for a more detailed discussion.

6. This section focuses on the financial community's perspective on the final agreement. I offer the views of various political actors in Chapter 5.

7. Amazingly, PSNH's share of Seabrook was valued at about $250 million in this transaction. The remainder of the price was for PSNH's nonnuclear assets.

8. Citicorp did own some PSNH third mortgage bonds.

9. In hindsight, this estimate proved to be mostly accurate, except that bondholders received a bit less than projected.

Chapter 5

1. Indeed, by 1981, 80 percent of the Shoreham–Wading River School District budget came from Shoreham's taxes. The district was considered one of the finest (and richest) on Long Island even with moderate property taxes on homeowners.

2. Unless specifically cited, information for this section comes passim from Bedford (1990), Chapters 1–3; Wasserman (1979); Grossman (1980); and Stever (1980). Also used were various issues of the *Boston Globe*, the *New York Times*, the *Wall Street Journal* and the *Manchester Union-Leader* between 1971 and 1985.

3. See note 2 for this section's sources.

4. New Hampshire Yankee interviews, June 1992.

5. New Hampshire Yankee interviews, June 1992.

6. Long Island Association interviews, October 1990.

7. Long Island Association interviews, October 1990.

8. Long Island Association interviews, October 1990.

9. New Hampshire Yankee interviews, June 1992.

10. Courts had ruled that the old township system was unconstitutional because it gave large towns control over small towns, thereby violating the "one person–one vote" premise. Similar legal challenges were made against Nassau's system, but there was never any ruling.

11. This section was augmented by interviews with LIA president James LaRocca and LIA board member John Marburger, the latter of which chaired the Marburger Commission in 1983 and was president of SUNY, Stony Brook.

12. Mr. LaRocca did not agree that there was any significant split within the LIA, just a gradual overall diminishing of support for Shoreham and for LILCO.

13. Much of the information on the BIA comes from New Hampshire Yankee interviews, June 1992.

14. New Hampshire Yankee interviews, June 1992.

15. Because New Hampshire is so small, it seems plausible to talk about

the entire state business community as "local." Almost all corporate activity is centered in the southern tier of New Hampshire.

16. New Hampshire Yankee interviews, June 1992.

17. New Hampshire Yankee interviews, June 1992.

Chapter 6

1. This dominant class is sometimes referred to as a "ruling class," "capitalist class," or "upper class."

REFERENCES

Ahearne, John.

   1987   Untitled essay. Resources 88: 5–9.

Alford, Robert, and Roger Friedland.

   1975   "Political Participation and Public Policy." Annual Review of Sociology 1: 429–479.

   1985   Powers of Theory. Cambridge: Cambridge University Press.

Balogh, Brian.

   1991   Chain Reaction. Cambridge: Cambridge University Press.

Bedford, Henry.

   1990   Seabrook Station: Citizen Politics and Nuclear Power. Amherst:
University of Massachusetts Press.

Berle, Adolph, and Gardiner Means.

   1932   The Modern Corporation and Private Property. New York: Har-
court, Brace, and World (reprinted 1968).

Burnham, James.

   1941   The Managerial Revolution. New York: John Day.

Camilleri, Joseph.

   1984   The State and Nuclear Power. Seattle: University of Washington
Press.

Campbell, John.

   1988   Collapse of an Industry. Ithaca, NY: Cornell University Press.

Chandler, Alfred.

   1977   The Visible Hand. Cambridge, MA: Harvard University Press.

Clarke, Lee.

   1985   "The Origins of Nuclear Power: A Case of Institutional Conflict."
Social Problems 32: 447–487.

Chen, Bernard.
  1990   The Nuclear Energy Option. New York: Plenum.
Commoner, Barry.
  1979   The Politics of Energy. New York: Knopf.
Condon, Edward.
  1958   "Bombs for Peace Hypocrisy." The Nation 187: 376.
Curtis, Richard, and Elizabeth Hogan.
  1980   Nuclear Lessons. Harrisburg, PA: Stackpole Books.
Cyert, Richard, and James March.
  1956   "Organizational Factors in the Theory of Oligopoly." Quarterly
         Journal of Economics 70: 44–64.
Daneke, Gregory, and George Lagassa (eds.).
  1980   Energy Policy and Public Administration. Lexington, MA: Lex-
         ington Books.
Delaney, Kevin.
  1992   Strategic Bankruptcy. Berkeley: University of California Press.
  1994   "The Organizational Construction of the Bottom Line." Social
         Problems 41(4): 497–518.
DelSesto, Stephen.
  1979   Science Politics and Controversy: Civilian Nuclear Power in the
         U.S., 1946–1974. Boulder, CO: Westview Press.
Dunbar, Leslie.
  1958   "The Controversy of Nuclear Power." Current History 34: 275–282.
Ebbin, Stephen, and Raphael Kasper.
  1974   Citizen Groups and the Nuclear Power Controversy. Cambridge,
         MA: MIT Press.
Eckstein, Rick, and Kevin Delaney.
  1993   "Institutional Investment in Troubled Corporations—A Sociolog-
         ical Analysis." American Journal of Economics and Sociology
         52(3): 291–306.
Edgar, Richard.
  1970   Urban Power and Social Welfare. Beverly Hills: Sage.
Ewen, Lynda.
  1978   Corporate Power and Urban Crisis in Detroit. Princeton, NJ:
         Princeton University Press.

Fitch, Robert, and Mary Oppenheimer.
  1970 "Who Rules the Corporation?" Socialist Revolution 1(4): 73–108;
        1(5): 61–114.
Ford, Daniel.
  1982 The Cult of the Atom. New York: Simon & Schuster.
Freitag, Peter.
  1983 "The Myth of Corporate Capture." Social Problems 30: 480–491.
Friedland, Roger.
  1983 Power and Crisis in the City: Corporations, Unions, and Urban
        Policy. New York: Schocken.
Friedland, Roger, and Donald Palmer.
  1984 "Park Place and Main Street: Business and the Urban Power Struc-
        ture." Annual Review of Sociology 10: 395–416.
Gamson, William, and Andre Modigliani.
  1989 "Media Discourse and Public Opinion on Nuclear Power." Amer-
        ican Journal of Sociology 95: 1–37.
Gandara, Arturo.
  1977 Electric Utility Decision Making and the Nuclear Option. Santa
        Monica, CA: The Rand Corp.
Glasberg, Davita S.
  1981 "Corporate Power and Control: The Case of Leasco Corporation
        and Chemical Bank." Social Problems 29: 104–116.
  1989 The Power of Collective Purse Strings: The Effect of Bank Hege-
        mony on Corporations and the State. Berkeley: University of Cal-
        ifornia Press.
Glasberg, Davita, and Michael Schwartz.
  1983 "Ownership and Control of Corporations." Annual Review of So-
        ciology 9: 311–332.
Goodman, Robert.
  1986 The Last Entrepreneurs. Boston: South End Press.
Gordon, Robert.
  1945 Business Leadership in the Large Corporation. Washington, DC:
        The Brookings Institute.
Gramsci, Antonio.
  1971 Selections from the Prison Notebooks. New York: International
        Publishers.

Green, Harold, and Alan Rosenthal.
    1963   Government of the Atom. New York: Atherton.

Grossman, Karl.
    1980   Cover Up: What You're Not Supposed to Know About Nuclear Power. Sagaponack, NY: Permanent Press.
    1990   "New, Improved Nukes." EXTRA! 3: 1.

Hardert, Ronald, et al.
    1989   "A Critical Theory Analysis of Nuclear Power: The Implications of Palo Verde Nuclear Generating Station." Humanity and Society 13: 165–186.

Hayes, Edward.
    1972   Power Structure and Urban Policy: Who Rules in Oakland? New York: McGraw-Hill.

Hertsgaard, Mark.
    1983   Nuclear Inc. New York: Pantheon.

Hodgins, Eric.
    1953   "The Atom: Ready for Business." Fortune 47: 142–145.

Holt, Donald.
    1979   "The Nuke that Became a Lethal Political Weapon." Fortune January 15: 74.

Jasper, James.
    1990   Nuclear Politics. Princeton, NJ: Princeton University Press.

Javits, Benjamin.
    1969   A Better World for All Through Democratic Ownership. New York: Crown.

JCAE (Joint Committee on Atomic Energy).
    1955   Hearings on Development Growth and State of the Atomic Energy Industry. Washington, DC: U.S. Government Printing Office.

Kaysen, Carl.
    1957   "The Social Significance of the Modern Corporation." American Economic Review 47: 311–319.

Koenig, Thomas, and Robert Gogel.
    1981   "Interlocking Directorates in a Social Network." American Journal of Economics and Sociology 40: 37–50.

Lamb, S., and J. Rappaport.
    1980   Municipal Bonds. New York: McGraw-Hill.

Loftus, Joseph.

1950   "Some Observations on the Economics of Atomic Power." Science Monthly 70: 396–401.

Logan, John, and Harvey Molotch.

1987   Urban Fortunes. Berkeley: University of California Press.

Long Island Lighting Company.

1984–1988 Annual Reports. Hicksville, NY.

Lonnroth, Mans, and William Walker.

1979   The Viability of the Civilian Nuclear Industry. New York: The Rockefeller Institute.

Lowenstein, Louis.

1988   What's Wrong with Wall Street? Reading, MA: Addison-Wesley.

Lukes, Steven.

1974   Power: A Radical View. London: Macmillan.

Marris, Robin.

1964   The Economic Theory of Managerial Capitalism. London: Macmillan.

Mazur, Alan.

1988   Mass Media Effects on Public Opinion of Nuclear Power in the U.S. Paper presentation. North Central Sociological Association. Pittsburgh, PA.

Mazuzan, George, and Samuel Walker.

1984   Controlling the Atom. Berkeley: University of California Press.

McCaffrey, David.

1991   The Politics of Nuclear Power: A History of the Shoreham Nuclear Plant. Boston: Kluwer.

McGuire, Patrick.

1986   "The Political Economy of the Rise of the Electrical Utility Industry in the U.S." Ph.D. dissertation. Department of Sociology, State University of New York at Stony Brook.

Mills, C. Wright.

1959   The Sociological Imagination. New York: Oxford University Press.

Mintz, Beth, and Michael Schwartz.

1985   The Power Structure of American Business. Chicago: University of Chicago Press.

1981   "Interlocking Directorates and Interest Group Formation." American Sociological Review 46: 851–869.

Mizruchi, Mark.
1982   The American Corporate Network. Beverly Hills: Sage.

Molotch, Harvey.
1976   "The City as a Growth Machine." American Journal of Sociology 82: 309–330.

O'Neill, Charles.
1959   "Investment Position of Electric Utilities." The Commercial and Financial Chronicle. Dec. 3: 2310.

Perucci, Robert, and Harry Potter (eds.).
1989   Networks of Power. New York: Aldine de Gruyter.

Ratcliff, Richard.
1979   "The Civic Involvement of Bankers." Social Problems 26: 298–313 (with M. E. Gallagher and K. Ratcliff).
1980   "Banks and Corporate Lending." American Sociological Review 45: 1553–1570.

Rockefeller, David.
1964   Creative Management in Banking. New York: McGraw-Hill.

Royce, Edward.
1993   The Origins of Southern Sharecropping. Philadelphia: Temple University Press.

Scaff, H. H.
1960   "Electrical Utilities Will Spend $52 Billion This Decade." The Commercial and Financial Chronicle. June 16: 2608.

Stever, James.
1980   Seabrook and the Nuclear Regulatory Commission. Hanover, NH: University of New Hampshire Press.

Temples, James.
1982   "The NRC and the Politics of Regulatory Reform." Public Administration Review 42: 355–362.

Useem, Michael.
1983   The Inner Circle: Large Corporations and Business Politics in the U.S. and Great Britain. Cambridge: Winthrop.

U.S. Senate.
1974   Financial Problems of the Electrical Utilities. Committee on In-

terior and Insular Affairs. Washington, DC: U.S. Government Printing Office.

Walker, Samuel.

1992   Containing the Atom. Berkeley: University of California Press.

Wasserman, Harvey.

1979   Energy Wars. Westport, CT: Lawrence Hill.

Weinberg, Alvin, et al. (eds.).

1985   The Nuclear Connection. New York: Paragon House.

White, Edward.

1961   "Compensating Balances and Regulatory Agencies." The Commercial and Financial Chronicle. June 15: 2619.

Whitt, J. Allen.

1982   The Dialectics of Power: Urban Elites and Mass Transportation. Princeton, NJ: Princeton University Press.

Wood, William.

1983   Nuclear Safety: Risks and Regulations. Washington, DC: American Enterprise Institute.

Zeitlin, Maurice.

1978   "Corporate Ownership and Control." American Journal of Sociology 81: 894–903.

Zorpette, Glenn, and Karen Fitzgerald.

1987   "The Shoreham Saga." IEEE Spectrum 24: 25–30.

Zweig, Michael.

1987   The Wine Industry and the Future of Agriculture on Long Island's North Fork. Stony Brook, NY: Institute for Social Analysis.

INDEX